Frontiers in Clinical Drug Research – Diabetes and Obesity

(*Volume 5*)

Edited by
Atta-ur-Rahman, *FRS*
Kings College, University of Cambridge, Cambridge, UK

Frontiers in Clinical Drug Research – Diabetes and Obesity

Volume # 5

Editor: Prof. Atta-ur-Rahman, *FRS*

ISSN (Online): 2352-3220

ISSN (Print): 2467-9607

ISBN (Online): 978-1-68108-753-5

ISBN (Print): 978-1-68108-754-2

Published by Bentham Science Publishers Pte. Ltd. Singapore. All Rights Reserved.

need for a court order if at any point you breach any terms of this License Agreement. In no event will any delay or failure by Bentham Science Publishers in enforcing your compliance with this License Agreement constitute a waiver of any of its rights.

3. You acknowledge that you have read this License Agreement, and agree to be bound by its terms and conditions. To the extent that any other terms and conditions presented on any website of Bentham Science Publishers conflict with, or are inconsistent with, the terms and conditions set out in this License Agreement, you acknowledge that the terms and conditions set out in this License Agreement shall prevail.

Bentham Science Publishers Pte. Ltd.
80 Robinson Road #02-00
Singapore 068898
Singapore
Email: subscriptions@benthamscience.net

CONTENTS

PREFACE

The fifth volume of Frontiers in Clinical Drug Research – Diabetes and Obesity comprises five comprehensive chapters discussing novel approaches to combat diabetes and obesity.

In the first chapter, Mandrioli *et al*, present the three most important classical neuroleptics (chlorpromazine, haloperidol and loxapine). The most important antipsychotics are individually analyzed in relation to their propensity to cause metabolic syndrome. In chapter 2 of the book, Sobrevia *et al* summarise some examples of the wide variety of protocols for insulin therapy and the potential consequences of this protocol on the foetoplacental unit and the neonate from women with Gestational diabetes mellitus (GDM).

Growing evidence suggests that hyperglycemia results in increased reactive oxygen species (ROS) production, leading to oxidative stress which affects and damages various tissues and organs. Oxidative stress results from an imbalance between ROS and antioxidants. Houreld and Rajendran highlight the understanding of oxidative stress-related mechanisms underlying the development of diabetes. Their review also elaborates on antioxidant therapy strategies to diminish oxidative stress and to treat diabetic associated complications.

Diabetes mellitus (DM) is a metabolic disorder which is the most alarming disease of the modern era. It occurs as a result of lack of insulin secretion or reduced insulin secretion or peripheral insulin resistance. In chapter 4, Anreddy *et al*. describe the issues concerned with the oral delivery of insulin and also discuss possible routes for the administration and use of Nanoparticles (NPs) for the best delivery of insulin.

In the last chapter of the book, Sharma *et al*. give comprehensive details about the merits and demerits of a class of drugs called Sodium-glucose co-transporter-2 (SGLT-2) inhibitors.

I owe special thanks to all the contributors for their valuable contributions in bringing together the fifth volume of this book series. I also thank the editorial staff of Bentham Science Publishers for their help and support.

Atta-ur-Rahman, *FRS*
Kings College,
University of Cambridge,
Cambridge,
UK

List of Contributors

Roberto Mandrioli	Department for Life Quality Studies, Alma Mater Studiorum – University of Bologna, Rimini, Italy
Michele Protti	Pharmaco-Toxicological Analysis Laboratory (PTA Lab), Department of Pharmacy and Biotechnology (FaBiT), Alma Mater Studiorum - University of Bologna, Bologna, Italy
Laura Mercolini	Pharmaco-Toxicological Analysis Laboratory (PTA Lab), Department of Pharmacy and Biotechnology (FaBiT), Alma Mater Studiorum - University of Bologna, Bologna, Italy
Mario Subiabre	Cellular and Molecular Physiology Laboratory (CMPL), Department of Obstetrics, Division of Obstetrics and Gynaecology, School of Medicine, Faculty of Medicine, Pontificia Universidad Católica de Chile, Santiago 8330024, Chile
Roberto Villalobos-Labra	Cellular and Molecular Physiology Laboratory (CMPL), Department of Obstetrics, Division of Obstetrics and Gynaecology, School of Medicine, Faculty of Medicine, Pontificia Universidad Católica de Chile, Santiago 8330024, Chile
Luis Silva	Cellular and Molecular Physiology Laboratory (CMPL), Department of Obstetrics, Division of Obstetrics and Gynaecology, School of Medicine, Faculty of Medicine, Pontificia Universidad Católica de Chile, Santiago 8330024, Chile Immunoendocrinology, Division of Medical Biology, Department of Pathology and Medical Biology, University of Groningen, University Medical Centre Groningen (UMCG), Groningen 9700 RB, The Netherlands
Fabián Pardo	Cellular and Molecular Physiology Laboratory (CMPL), Department of Obstetrics, Division of Obstetrics and Gynaecology, School of Medicine, Faculty of Medicine, Pontificia Universidad Católica de Chile, Santiago 8330024, Chile Metabolic Diseases Research Laboratory, Center of Research, Development, and Innovation in Health - Aconcagua Valley, San Felipe Campus, School of Medicine, Faculty of Medicine, Universidad de Valparaíso,, 2172972 San Felipe, Chile
Luis Sobrevia	Cellular and Molecular Physiology Laboratory (CMPL), Department of Obstetrics, Division of Obstetrics and Gynaecology, School of Medicine, Faculty of Medicine, Pontificia Universidad Católica de Chile, Santiago 8330024, Chile Department of Physiology, Faculty of Pharmacy, Universidad de Sevilla, Seville E-41012, Spain University of Queensland Centre for Clinical Research (UQCCR), Faculty of Medicine and Biomedical Science, University of Queensland, Herston, QLD 4029, Queensland, Australia
Nicolette Nadene Houreld	Laser Research Centre, Faculty of Health Sciences, University of Johannesburg, Johannesburg, 2028, South Africa
Naresh Kumar Rajendran	Laser Research Centre, Faculty of Health Sciences, University of Johannesburg, Johannesburg, 2028, South Africa

Radhika Tippani   Department of Biotechnology, Kakatiya University, Warangal, 506009, Telangana State, India

Rama Narsimha Reddy Anreddy   Department of Pharmacology, Jyothishmathi Institute of Pharmaceutical Sciences, Ramakrishna Colony, Thimmapur, Karimnagar 505481, Telangana State, India

Mahendar Porika   Department of Biotechnology, Kakatiya University, Warangal, 506009, Telangana State, India

Siddhartha Dutta   All India Institute of Medical Sciences, Jodhpur, Rajasthan, India

Pramod Kumar Sharma   All India Institute of Medical Sciences, Jodhpur, Rajasthan, India

Arup Kumar Misra   All India Institute of Medical Sciences, Jodhpur, Rajasthan, India

Metabolic Syndrome in Schizophrenia: Focus on the Role of Antipsychotic Medications and Indications for Therapeutic Drug Monitoring (TDM) Methods

Roberto Mandrioli[1,*], Michele Protti[2] and Laura Mercolini[2]

[1] *Department for Life Quality Studies, Alma Mater Studiorum – University of Bologna, Rimini, Italy*

[2] *Pharmaco-Toxicological Analysis Laboratory (PTA Lab), Department of Pharmacy and Biotechnology (FaBiT), Alma Mater Studiorum - University of Bologna, Bologna, Italy*

Abstract: Metabolic syndrome is a complex pathology characterized by imbalances in lipid and glucose metabolism and weight gain, and consequently by an increase in the incidence of type II diabetes and cardiovascular disease. Metabolic syndrome is rapidly becoming one of the most important side effects of treatment with modern "atypical" antipsychotic agents, probably due to their specific mechanisms of action. Although the most recent members of this class (aripiprazole, asenapine, ziprasidone) seem to produce a reduce incidence of metabolic syndrome, the problem is far from being resolved. In this chapter, the three most important classical neuroleptics (chlorpromazine, haloperidol and loxapine) and the most important atypical antipsychotics will be individually analyzed in relation to their propensity to cause metabolic syndrome. The most reliable, current data will be presented, also in the perspective of possible interventions to mitigate metabolic imbalances, comparative studies, switching studies and augmentation strategies. An important strategy for metabolic syndrome prevention could also be the performing of an accurate therapeutic drug monitoring (TDM). Thus, an up-to-date overview will also be presented of recent and significant analytical methods for the determination of the drugs of interest and their main metabolites in human biological fluids.

Keywords: Antipsychotic drugs, Diabetes, Hypercholesterolemia, Hypertrigly-ceridemia, Metabolic syndrome (MetSyn), Obesity, Pharmacotherapy, Polypharmacy, Therapeutic Drug Monitoring (TDM).

---

* **Corresponding author Roberto Mandrioli:** Department for Life Quality Studies, Alma Mater Studiorum – University of Bologna, Corso d'Augusto 237, I-47921 Rimini, Italy; E-mail: roberto.mandrioli@unibo.it

Atta-ur-Rahman (Ed.)

INTRODUCTION

Psychiatric disorders are currently one of the main causes of disability and years lost to illness all over the world: According to recent World Health Organization (WHO) statistics, almost 30% of people have experienced a common mental disorder at some time during their lifetimes [1]. Psychiatric disorders often take a heavy toll on patients' well-being also from a purely physical point of view. Since they often cannot take care of themselves optimally, comorbidity is very frequent, and naturally it increases with age. On top of all of this, one must remember that pharmacological therapy, albeit one of the most effective forms of treatment, is also in itself a potential source of iatrogenic effects. According to some studies, it is estimated that up to 74% of patients discontinue the medication before 18 months [2], and between 10 and 20% are forced to interrupt the pharmacologic treatment due to side or toxic effects [2, 3]. Two major factors contribute to exacerbate this situation: polypharmacy and lifelong chronic therapy. Psychiatric patients are often subjected to polypharmacy due to the inherent difficulty in controlling the disorder's multifaceted symptoms, and also due to the relatively low rate of success of the therapy. It is estimated that between 20 and 60% of schizophrenic patients are currently treated with two or more drugs for their illness [4]. Since schizophrenia is a very complex, articulated syndrome, different drugs are usually able to control just a part, or some aspects of the overall symptoms; hence the need for polypharmacy. Of course, this situation becomes even more complicated and worrying for elderly patients, who are usually subjected to pharmacological therapies for other severe illnesses as well as for the psychiatric disorder. Regarding the therapy duration, psychiatric disorders are eminently chronic. What is worse, current treatments are not resolutive of the underlying problem, nor an etiologic agent is currently known. As a consequence, lifelong treatment with one or more drugs is relatively frequent [5]; periods of remission, followed by one or more relapses, are quite common as well [6]. Minimum therapy duration is measured in months or years, not in weeks.

Many side effects of antipsychotic drugs are well known and readily taken care of during the treatment (*e.g.*, extrapyramidal symptoms); however, a few are still kept in the background and not completely acknowledged or understood. Perhaps the most important of these latter effects is metabolic syndrome (MetSyn). MetSyn is a chronic multifactorial disease related to several conditions that have the common trait of increasing the probability of a cardiovascular event. The most important conditions involved in MetSyn are diabetes, central or visceral obesity, hypercholesterolemia, hypertriglyceridemia and hypertension. The red line that connects these conditions is a metabolism imbalance associated with an insulin resistance state and an activation of the sympathetic nervous system [7, 8]. Since cardiovascular events are the leading cause of death and disability in the world's population, it is easily understood how MetSyn could well be one of the most important diseases to attract the attention of clinicians and researchers alike. Criteria for the differential diagnosis of MetSyn, both epidemiologically and in clinical practice, have been laid out in 1999 by the WHO [9], in 2001 by the American Medical Association (in the National Cholesterol Education Program – NCEP – III definitions) [10] and in 2006 by the International Diabetes Federation [11].

A 2004 study based on prospective European cohort studies (more than 11,000 subjects) found a MetSyn prevalence of 15.7% for men and 14.2% for women [12]; since this study excluded diabetic patients, its results are probably underestimating the prevalence. Other papers underline that, due to the multiplicity of symptoms and measurement methods, it is very difficult to obtain reliable prevalence data and to compare different populations [13]. For example, in 2003 a discrepancy of 13% (13.6% *vs.* 26.6%) between the WHO and NCEP-III criteria was reported for the prevalence of MetSyn in Mexican subjects; even excluding diabetic patients, this discrepancy was largely maintained (9.2% *vs.* 21.4%, respectively). However, most studies agree that MetSyn prevalence increases with age: In the year 2000 in the USA, MetSyn prevalence ranged from 6.7% for patients aged 20-29 years to 42% for those aged more than 70 years. The age-adjusted prevalence was 23.7% [14].

Metabolic Syndrome in Psychotic Patients

Obviously, most patients suffering from schizophrenia are treated with some form of pharmacotherapy. As a consequence, it is difficult to separate the direct effect of the disorder on patients' health from that of the medications. Anyway, recent studies on the prevalence of MetSyn in schizophrenic patients are really alarming. About 55% of 252 Dutch patients were found to meet IDF criteria for MetSyn [15]; in a 10-year retrospective study on 174 Malaysian patients, 36% developed metabolic syndrome, 23% were hypertensive and 28% were diabetic, but 100% of them had significantly increased weight, body mass index (BMI), fasting blood sugar and blood pressure [16]. These prevalence data are higher than those of the general population, and should be an important source of alarm for clinicians during the therapy. Nowadays, it is widely acknowledged that antipsychotic therapy can have important, negative effects on the onset of MetSyn.

The Role of Therapeutic Drug Monitoring in Metabolic Syndrome Prevention and Treatment

The frequency and severity of MetSyn can often be correlated with chemical-clinical parameters, in particular with the plasma levels of the antipsychotic medication or some of its metabolites. This positive correlation between plasma levels and MetSyn has been reported, in fact, for clozapine [17]; an inverse correlation has been observed for its major metabolite norclozapine [18]. A correlation between risperidone plasma levels and metabolic side effects has been also shown in children and adolescents [19]. During quetiapine treatment, the amount of weight gain has been found to increase with higher drug doses [20]; although not investigated in this study, plasma quetiapine levels are conceivably correlated to this finding. In nonsmoking patients, it has been found that plasma levels of the main olanzapine metabolite (*N*-desmethylolanzapine) are negatively correlated with blood glucose and insulin concentrations [21]. More recently, another study has concluded that the plasma concentration / dose (C/D) ratio of *N*-desmethylolanzapine correlates negatively with weight, BMI, waist circumference, and C-peptide level. Moreover, an olanzapine/N-desm--thylolanzapine plasma level ratio higher than 6 was positively correlated to MetSyn [22].

All of these results are strong clues for the possibility that metabolic side effects during antipsychotic therapy are subject to chemical-clinical correlations (CCCs), *i.e.*, relatively strict correlations between drug and metabolite plasma levels and therapeutic, side and toxic effects. For these reasons, therapeutic drug monitoring (TDM) can be advisable during antipsychotic therapy, and specifically in the case of atypical antipsychotics.

According to the workgroup on TDM (Arbeitsgruppe "Therapeutisches Drug Monitoring") of the German Arbeitsgemeinschaft für Neuropsychopharmakologie und Pharmakopsychiatrie (AGNP, Association for Neuro-Psychopharmacology and Pharmacopsychiatry), "TDM uses the quantification of drug concentrations in blood plasma to titrate the dose of individual patients so that a drug concentration associated with the highest possible probability of response and tolerability and a low risk of toxicity can be obtained. Moreover, TDM has the possible and widely unexploited potential to improve cost-effectiveness of psychopharmacotherapy" [23].

In fact, the determination of drug (and metabolites) plasma levels gives the clinician "objective" tools for a better assessment of how the drug behaves within the individual patient's body. It is also useful in determining patient compliance, since it can point out to either missed drug doses (resulting in very low or absent drug in plasma) and drug overdoses (resulting in very high plasma levels) [24]. In this way, the pharmacological treatment efficacy can be assessed in each patient with both clinical and chemical means, reducing uncertainty and helping the clinician in tailoring the therapy to the need of the patient regarding dose, schedule and drug choice. This in turn can lead to lower incidence of side and toxic effects and lower hospitalization rates, and thus to better quality of life for patients and relatives, and lower total expenses for public health and health care services [25].

One of the main tenets of an effective routine application of TDM is the existence and knowledge of CCCs [26, 27], also including the correlation between administered dose and plasma levels. However, some form of monitoring can also be experimental, using the routine determination of drug plasma levels with the aim of establishing CCCs in those cases when they are not certain or not known yet. TDM for the control of metabolic side effects would fall within this provision, since accepted relationships between plasma levels and MetSyn symptoms or parameters are still not known.

For a considerable number of psychopharmacologic compounds, the quantitation of the medication plasma concentrations has become clinical routine for dose adjustment: evidence of the benefits of TDM has been produced for tricyclic antidepressants [28], (with dedicated analytical methods [29]), for more recent selective serotonin reuptake inhibitors (SSRIs) [27] and for other second-generation antidepressants [30]. Similarly, the practice is becoming ever more frequent for a number of mood stabilizing drugs [23, 31, 32] and, more slowly and not without resistance, for anxiolytics and hypnotics [33 - 35]. However, the therapeutic class that perhaps has the best perspectives in terms of TDM effectiveness is the antipsychotic agent class [36]; incidentally, they are probably the CNS drugs with the longest story of successful TDM application, apart from anti-seizure drugs.

In fact, for some aspects antipsychotic drugs are ideal candidates for TDM: they are usually part of a chronic therapy, have a relatively high incidence of side effects and a relatively high discontinuation rate due to side/toxic effects; polypharmacy is frequent, with increased risks of drug interactions; interindividual variability of plasma levels is high; and finally, patients with schizophrenia are often wary of any pharmacological intervention and for this reason compliance is typically low [37].

The suggested monitoring frequency for psychotropic medications is usually one sampling every 3-6 months [23], but this is not absolutely fixed. Practically, the monitoring interval should be established based on the expected drug/metabolite metabolism and pharmacokinetics (to be adjusted according to the patient's physiology), previous monitoring results and therapy efficacy or lack thereof. Other contingent or episodic factors can warrant more frequent or supplemental samplings and analyses: for example, suspected lack of compliance or overdose; recent therapeutic regimen (dose, drug) changes; side or toxic effect manifestations or other sudden changes in the patient's health conditions;

concomitant pathologies, pharmacological treatments or changes in the diet or life style (*e.g.*, smoking cessation). Blood sampling is usually carried out at the trough level, which is often in the morning, just before the first daily drug dose; in this way, measurements are more stable and reproducible. In the last few years, a trend can be seen toward the increasing use for TDM of alternative matrices (saliva, hair, nails, sweat) and of micromatrices – such as dried blood spots (DBS) [38], dried plasma spots (DPS), volumetric absorption microsampling (VAMS) – or miniaturized sample preparation procedures – such as microextraction by packed sorbent (MEPS), disposable pipette extraction (DPX), solid phase microextraction (SPME) [39, 40]. While currently these approaches have only found niche applications, their undeniable advantages in terms of low invasiveness, speed, ease of storage and analyte stability will probably bring them to the forefront of TDM in the next years.

Of course, TDM is not, and is not claimed to be, the "definitive weapon" in the psychiatrist's arsenal. It should be noted that TDM, compared to traditional no-monitoring therapy, surely represents an additional upfront expense for health care services or for patients. However, due to the low sampling frequency, even expensive analyses represent a very limited cost increase over the personnel and drug purchase costs; a cost increase that is more than compensated by the corresponding decrease in expenses for side and toxic effect treatment and for supplemental hospitalizations.

CLASSICAL NEUROLEPTICS AND METABOLIC SYNDROME

Classical (first-generation) neuroleptics such as phenothiazines, butyrophenones and thioxanthenes are not considered prone to cause weight gain, insulin resistance and in general MetSyn. In some cases, switching from a second-generation antipsychotic to a first-generation one can even be considered a good strategy to counter the severe effects of MetSyn. However, several differences exist within this class, and each drug should be evaluated on its own merits in relation to MetSyn. In any case, due to their favorable metabolic profile, it is generally acknowledged that there is no need to carry out a TDM schedule to prevent or minimize this kind of side effects. Thus, no analytical methods are described for these drugs.

Chlorpromazine

Although it is well known that chlorpromazine can cause weight gain and metabolic abnormalities, studies on its metabolic effects are not numerous and relatively old [41, 42]. A 2005 study on sexual disturbances in male humans found increased levels of blood glucose, LDL and apolipoprotein B [43] in patients treated with chlorpromazine. However, only 17 patients were treated with this drug in the study. A study on Wistar rats has suggested that chlorpromazine could contribute to inducing metabolic syndrome by acting on blood glucose levels and insulin resistance [44]. This activity, in turn, would be due to the drug's induction of adipocytokines, in particular adiponectin, whose levels were elevated in rats treated with chlorpromazine.

Haloperidol

Studies regarding haloperidol have been published even recently. A quite large, open-label study on 300 out-patients with a long history of schizophrenia established that no difference in the metabolic syndrome rate existed between patients treated with haloperidol, olanzapine and aripiprazole [45]. Overall, rates of metabolic syndrome onset were quite high for all tested drugs, reaching 37% in the haloperidol group. Conversely, a study on psychotic patients experiencing acute relapse found no influence by haloperidol on body weight, or on glucose, triglyceride and total cholesterol levels, after 6 weeks of treatment and also when compared to placebo [46]. However, this study failed to separate haloperidol (which was included for sensitivity) from placebo in most indices, resulting in a failed study. Emsley *et al.* [47] have found elevated levels of glycated hemoglobin (HbA1c) in patients treated with haloperidol, although no impairment in glucose tolerance was found in another study [48] after 8 weeks of treatment.

Other studies found that switching from haloperidol to ziprasidone can cause an increase in body weight [49].

A study on rats has found contrasting evidence on haloperidol effects on parameters related to MetSyn: in this study, rats treated with haloperidol experienced significant weight loss, as well as no significant effects on percentage of fat tissue and adiponectin, leptin, ghrelin, fasting glucose and insulin levels. HbA1c even decreased in female rats, while cholesterol decreased in male rats [50].

Loxapine

Just one study has been carried out on loxapine in the context of MetSyn [51], but its results seem to be encouraging. In fact, loxapine addition was tested in 15 patients with autism spectrum disorder (ASD) who were experiencing weight gain due to antipsychotic treatment. All of them were receiving atypical antipsychotics (risperidone, aripiprazole, quetiapine, olanzapine, ziprasidone), except one who was being treated with chlorpromazine. After loxapine addition, the previous antipsychotic was tapered or discontinued, and the patients experienced body weight loss, reduced body mass index (BMI) and decreased blood triglycerides.

ATYPICAL ANTIPSYCHOTICS AND METABOLIC SYNDROME

Differently from classical neuroleptics, atypical (or second-generation) antipsychotics include MetSyn among their most important side effects, one that can often lead to therapy discontinuation. This fact is often correlated to the different receptor affinity profile of atypical antipsychotics, which interact with both dopaminergic (D) and serotonergic (5-HT) receptors, whereas classical neuroleptics act exclusively at D receptors. Since 5-HT receptor activity is clearly related to weight gain and other metabolic imbalances, its further connection to MetSyn is easily understandable.

Amisulpride

Although amisulpride (4-amino-*N*-[[(1-ethylpyrrolidin-2-yl]methyl]-5-ethyl - sulfonyl-2-methoxybenzamide, Fig. **1**) is usually considered an atypical antipsychotic, its receptor binding profile is quite different from that of other atypical antipsychotics. In fact, it lacks affinity toward 5-HT receptors, which are usually considered the main receptors involved in weight gain and other manifestations of MetSyn. As a consequence, amisulpride is associated with a low risk of metabolic syndrome [52].

The risk of weight gain was evaluated as lower than that of risperidone in a comparative study [53] and in a six-month double-blind study [54]. An eight-week study [55] and a six-month study [56], both double-blind, established that the level of weight gain is higher for olanzapine than for amisulpride. A meta-analysis of randomized double-blind studies also concluded that the pharmacological profile of amisulpride is favorable regarding metabolic syndrome [57].

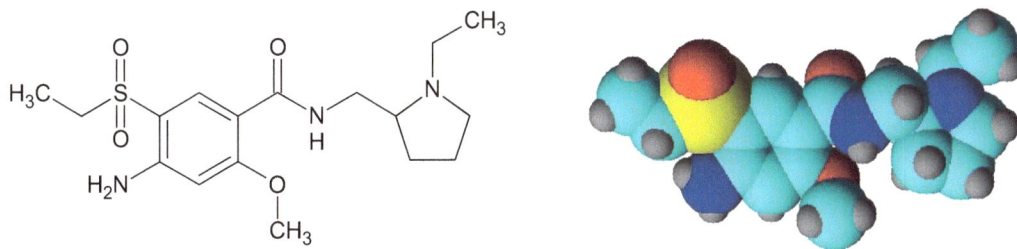

Fig. (1). Conventional chemical structure and 3D-spacefill structure of amisulpride.

However, recently a few reports have begun to surface of patients who experienced body weight gain [58], possibly associated with hypertriglyceridemia, lowered HDL, and hyperprolactinemia [59]. In a 12-week comparative study between amisulpride and ziprasidone on negative symptoms of schizophrenia [60], the former caused a larger rate of significant body weight increase (18% *vs.* 9%) and a similar rate of weight decrease (5% *vs.* 7%) in comparison to the latter.

Analytical Methods Suitable for TDM

Some methods have been published on amisulpride analysis in biological fluids: most of them regard the simultaneous determination of several drugs [61 - 63], usually for screening, toxicological and forensic purposes. Two recent multianalyte methods that include amisulpride have been published, one based on liquid chromatography with tandem mass detection (LC-MS/MS, with QTrap detection) and one based on microflow liquid chromatography (MFLC) - MS/MS with TQ detection; both have similar performances [64].

Some high-performance liquid chromatographic (HPLC) methods are also available, with ultraviolet (UV) [65 - 68], fluorescence (FL) [69] or MS/MS [70, 71] detection, for the determination of amisulpride alone (or with other benzamides) in biological fluids. Recently, an online solid phase extraction (SPE) - HPLC - photodiode array (PDA) detection method has been published, which uses a restricted access material (RAM) column to efficiently separate amisulpride from plasma macromolecules [72].

An SPE - gaschromatography (GC) - MS [73] and a voltammetric method [74] are available as well. The enantioselective analysis of amisulpride has been also carried out, by an HPLC method with combined UV and FL detection [75].

Aripiprazole

Aripiprazole (7-[4-[4-(2,3-dichlorophenyl)piperazin-1-yl]butoxy]-3,4-dihydro-1*H*-quinolin-2-one, Fig. **2**) is currently considered one of the atypical antipsychotics with the least pronounced tendency to cause MetSyn. The prevalence of MetSyn in patients who were initially stabilized with open-label aripiprazole for up to 18 weeks and then randomized to aripiprazole *versus* placebo for a maintenance phase of 26 weeks showed an initial increase, followed by a tendency to normalization [76]. In fact, patients remaining on aripiprazole for up to 26 weeks during the maintenance phase had a prevalence of MetSyn, and of any of its individual components, not significantly different from placebo [77]. In the treatment of schizoaffective disorder, there were no statistically significant differences at the 4-week endpoint between aripiprazole and placebo in the mean change in weight, glucose, or total cholesterol of 179 patients [78]. Nonetheless, it has been reported that weight gain is among the most frequent side effects in children and adolescents treated with aripiprazole for non-psychotic disorders (but not in those treated for psychotic disorders) [79]. Weight gain was one of the most significant side effects (44% frequency) also for young patients who were treated with the drug for first-episode schizophrenia in a 1-year, open-label, naturalistic outcome study [80]. In pediatric patients, weight gain with aripiprazole were more frequent if the patients were antipsychotic-naïve, younger and already overweight at baseline; overall, no differences in metabolic measures were observed in comparison to placebo [81]. No significant differences in body weight changes between aripiprazole and placebo were also observed in a classic 6-week, randomized, double-blind, placebo-controlled study [82]. Likewise, no significant differences in weight gain between placebo and aripiprazole were found in another classic study on patients with rapid-cycling bipolar disorder [83].

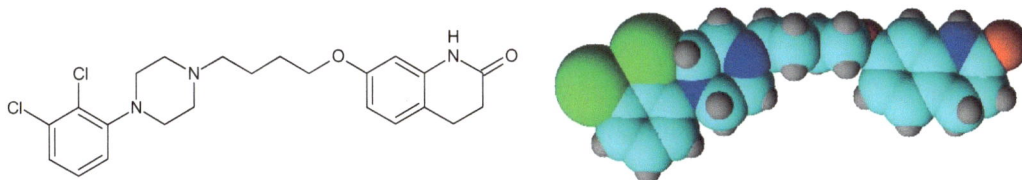

Fig. (2). Conventional chemical structure and 3D-spacefill structure of aripiprazole.

Children and adolescents with Tourette's disorder experienced significant increases in body weight, BMI and waist circumference during aripiprazole treatment, as compared to placebo [84]; comparison with pimozide showed that aripiprazole significantly increased cholesterolemia, while pimozide significantly increased glycemia [85].

Comparative Studies

A meta-analysis of randomized, controlled trials in Japanese patients has concluded that the metabolic profile of aripiprazole is more favorable than that of pooled antipsychotics, with advantages in terms of weight and total cholesterol and triglyceride levels [86]. The double-blind, randomized tolerability and efficacy study comparing quetiapine extended-release (ER) to aripiprazole in 113 children and adolescents with first-episode psychosis has confirmed that aripiprazole causes less metabolic effects than quetiapine [87]. A 1-year, randomized trial on 300 outpatients found no difference in MetSyn rates between olanzapine and aripiprazole [45]. However, a pooled analysis of three randomized clinical trials observed a significantly lower risk of developing MetSyn with aripiprazole compared with olanzapine for schizophrenic and bipolar patients after 1 year of therapy [88]. Both the rate and incidence of MetSyn were lower for aripiprazole than for olanzapine in a meta-analysis of four double-blind, randomized, controlled clinical trials [89]. In a classic randomized, double-blind study, statistically significant differences in weight change were observed between olanzapine and aripiprazole from week 1 to the endpoint (week 26). Fasting plasma levels of total cholesterol, HDL, and triglycerides were also significantly worse among patients treated with olanzapine [90].

Ethnicity can also play a role in the onset of MetSyn during antipsychotic treatment. In a post-hoc analysis of data from a 26-week, double-blind, randomized trial of aripiprazole and olanzapine, olanzapine significantly worsened all metabolic parameters except HDL and fasting glucose in Caucasian subjects. This was significantly different from aripiprazole for every outcome except fasting glucose. In the black/Hispanic cohort, olanzapine treatment resulted in adverse metabolic outcomes, and these changes were significantly different from aripiprazole for adiposity, total cholesterol, and non-HDL cholesterol. Aripiprazole significantly decreased the odds of endpoint MetSyn compared with olanzapine for all subjects and for the Caucasian cohort, but not for the black/Hispanic cohort. Within the aripiprazole group, Caucasian subjects had significantly lower risk for MetSyn, but there was no significant difference in subjects exposed to olanzapine.

Comparison with risperidone in a meta-analysis on autistic children found no difference in terms of weight gain [91]. A 52-week, follow-up, comparative study on the efficacy and tolerability of paliperidone (extended release), aripiprazole and ziprasidone on 203 patients with first-episode schizophrenia has found that efficacy and metabolic disruption are approximately inversely proportional, with paliperidone having the highest efficacy and the highest propensity to impair lipid metabolism (but not glucose metabolism), and aripiprazole having negative effects on glucose metabolism, body weight and BMI; however, it significantly decreased triglyceride levels [92]. A comparison of aripiprazole, olanzapine and risperidone in Korean patients with schizophrenia showed that the patients taking aripiprazole had a significantly lower risk of developing MetSyn. However, logistic regression identified just age and gender, and not the type of antipsychotic, as significantly correlated to MetSyn [93].

In the treatment of bipolar disorder, changes observed in metabolic parameters were comparably modest and similar in patients treated for up to 1 year with either lithium or aripiprazole [94].

Switching Studies

In 62 psychotic patients, switching from olanzapine to aripiprazole led to continuous improvements in waist circumference, blood pressure, triglyceride levels, blood glucose, and HDL. At the end of the 24-week, randomized, open-label study, 100% of the patients taking olanzapine met the criteria for MetSyn, as compared to 42.8% for those taking aripiprazole [95]. Switching from olanzapine, quetiapine or risperidone to aripiprazole seems to reduce coronary heart disease risk, but without any significant statistical difference in MetSyn prevalence or severity [96]. In a case series, some authors have observed that switching to aripiprazole from other antipsychotics after recent onset of MetSyn or type 2 diabetes can revert these pathologies to normality: all 7 cases of diabetes, and 50% of MetSyn cases, were reversed at 3 months follow-up. Moreover, there was a significant decrease in body weight, BMI, waist circumference, fasting glucose, fasting insulin, insulin resistance index, and serum lipid levels [97]. However, another study on 15 patients found that switching to aripiprazole did not alter weight or metabolic outcomes (fasting glucose, insulin resistance, and lipid concentrations) in a patient cohort who were 73% insulin resistant and 47% diabetic glucose tolerant at baseline [98].

In an 8-week, open, flexible-dose, trial on thirty-three schizophrenia patients who switched from other antipsychotics to aripiprazole, significant decreases in weight, waist circumference, LDL, glycemia and triglycerides were observed when switching from olanzapine. Decreases in the same parameters were not significant when switching from other antipsychotics [99]. In other studies still, switching from olanzapine produced a decrease of body weight and fasting triglyceride levels [100].

On the contrary, switching from aripiprazole to ziprasidone significantly decreased body weight, waist and hip circumferences and fasting blood glucose in a 12-week, open-label study on 19 patients [101].

Augmentation Strategies

A meta-analysis of 55 randomized, controlled, blind and non-blind trials has observed that adjunct aripiprazole is better than both placebo and open antipsychotic treatment for body weight and BMI [102]. In an 8-week open-label trial, aripiprazole augmentation seemed to bring about significant metabolic advantages for olanzapine-treated patients, but not for those treated with amisulpride, quetiapine or risperidone [103]. Augmenting risperidone with aripiprazole has produced lower LDL levels and stopped the weight gain associated with risperidone treatment in an 8-week, randomized, open-label trial on 130 patients [104]. Augmentation of clozapine with aripiprazole does not seem to provide remarkable benefits for MetSyn, although some body weight and plasma LDL reductions have been observed [105, 106]. However, aripiprazole augmentation could lower diastolic blood pressure [107].

Analytical Methods Suitable for TDM

From an analytical point of view, aripiprazole is one of the most intensely studied atypical antipsychotics in recent years. A few methods describe the analysis of aripiprazole (without any metabolite) in human plasma or serum [61, 108 - 113]. The analysis is mainly carried out by means of HPLC-UV [108], LC-MS/MS [61, 109 - 111, 113], UHPLC-MS/MS [112]. The sample pre-treatment can be based on liquid-liquid extraction (LLE) [109 - 111], plasma protein precipitation (PPP) [61] or SPE [112, 113]; one paper reports the use of a column switching system [108]. A single luminescence method is available [114], which is based on the fact that the weak chemiluminescence produced by the reaction of tris(1,10-phenanthroline)-Ru(II) with acidic Ce(IV) is enhanced in the presence of aripiprazole. In this case, plasma samples were subjected to PPP before analysis, which was carried out by the standard addition method.

Other methods also include one or more metabolites: dehydroaripiprazole [115 - 120], dehydroaripiprazole, 1-(2,3-dichlorophenyl)piperazine and 3,4-dihydro-7-(3′-carboxy)propoxy -2(1*H*)quinolinone [121]. These methods are based on LC-MS/MS [115, 116, 118, 119, 121], GC-MS [117] or HPLC-UV [120] after 96-well microelution plate SPE [118], SPE [117], LLE [116, 120] or PPP [115, 119, 121].

Recently, an LC-MS/MS (ESI-TQ) method has been published for the determination of the prodrug aripiprazole lauroxil and three of its main metabolites, which of course include aripiprazole as well as *N*-hydroxymethylaripiprazole and dehydroaripiprazole. SPE was used as the pre-treatment procedure for all analytes except *N*-hydroxymethylaripiprazole, for whom PPP was used [122].

Asenapine

Despite its intense affinity towards 5-HT receptors, and in particular 5-HT$_7$ ones, asenapine (5-chloro-2,3,3a,12b-tetrahydro-2-methyl-1*H*-dibenz(2,3-6,7)oxepi-o-(4,5-c)pyrrole, Fig. **3**) is currently considered one of the atypical antipsychotics with the least chance to produce MetSyn [123, 124]. Comparative studies *versus* risperidone [125] and *versus* olanzapine [126] in schizophrenia have found a significantly lower tendency of asenapine to cause weight gain, and lower mean weight gains (1.6 kg *vs.* 0.47 kg for risperidone, 5.5. kg *vs.* 1.6 kg for olanzapine); asenapine treatment was associated to slight decreases in triglycerides and cholesterol as well, while olanzapine was associated with increases in both parameters [126]. Comparative trials *versus* olanzapine in acute mania have reported similar results [127, 128], but with an increased tendency of asenapine to cause elevated glycemia.

Asenapine has even been used, at least episodically, to revert the weight gain caused by olanzapine therapy [129]; this use finds its rationale in systematic reviews and meta-analyses, both comparative [130, 131] and non-comparative [132, 133] ones.

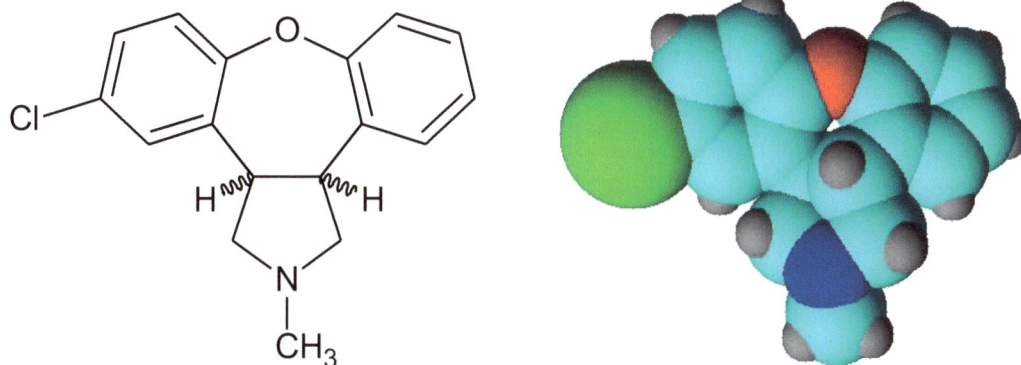

Fig. (3). Conventional chemical structure and 3D-spacefill structure of asenapine.

However, not all reports are equally optimistic regarding asenapine benign role in MetSyn: a 50-week open-label study on pediatric patients has reported that the incidence of clinically significant (>7%) weight gain during asenapine treatment can reach almost 35% [134]. Non-comparative trials have also found that asenapine has a higher rate of weight gain when compared to placebo (3.7% *vs.* 0.5% [135]).

Analytical Methods Suitable for TDM

A few methods are available for the analysis of asenapine in human biological fluids for TDM purposes. Two methods based on LC-MS/MS have been published [136, 137], both developed by the drug manufacturer to quantify asenapine and its main metabolites in blood [136] or urine [137] during the clinical trials. The two methods are very similar, but use two different sample pre-treatment procedures for plasma: automated SPE in 96-well plates for most metabolites and automated online SPE for the glucuronide. In urine, a single sample pre-treatment procedure based on automated online SPE was used. Two multi-analyte LC-MS/MS methods also include asenapine among their analytes, but without any metabolite [138, 139]. Another LC-MS/MS method includes two inactive asenapine metabolites, as well as the parent drug [140].

A paper has been published, describing a GC-MS method for the analysis of asenapine (no metabolites) in different biological tissues for forensic purposes [141], after liquid-liquid extraction and without prior derivatization. The lack of metabolite inclusion and the relatively low sensitivity make this method less than ideal for TDM purposes and more suitable for toxicological analyses. In the last few years, applications to alternative biological matrices have started to appear: for example, serum, urine and cerebrospinal fluid using ultrahigh performance liquid chromatography (UHPLC) - PDA [142], or hair and nails using HPLC-UV [143]. Although not immediately applicable to TDM, the use of these alternative matrices can be preliminary steps towards future, possible innovative practices for less invasive and more attractive sampling procedures.

Finally, a novel method for the enantioseparation and analysis of asenapine enantiomers in innovative micromatrices (DBS, DPS, blood VAMS, plasma VAMS) has been published [144]. The numerous advantages of micromatrices in comparison to classical blood/plasma/serum samples [40, 145], as well as its application to chiral analysis, make this method particularly promising for TDM purposes.

Clozapine

Clozapine (3-chloro-6-(4-methylpiperazin-1-yl)-5H-benzo[b] [1, 4] benzodia-zepine, Fig. (**4**)) is commonly regarded as the first atypical antipsychotic (it has been introduced onto the market in 1972). As such, it is probably the most well-known and most studied member of this class. Its tendency to cause metabolic effects is equally well-documented, to the point that some authors have recently proposed to begin metformin treatment simultaneously with clozapine treatment to reduce glucose level and body weight increases (project CoMET [146]). Other authors have also observed that some metabolic dysfunctions, and namely weight gain and hypertriglyceridemia, are strictly predictive of clinical response to clozapine; they have hypothesized that the very mechanism of action of clozapine could be linked to serum lipid changes [147]. As a result, it is nearly impossible (and largely useless) to discuss all papers published on the subject. In this discussion, only the most recent, important data will be provided, on studies regarding MetSyn during clozapine therapy.

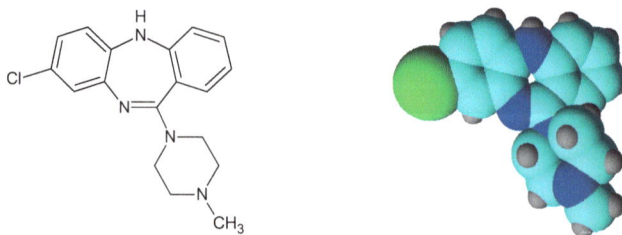

Fig. (4). Conventional chemical structure and 3D-spacefill structure of clozapine.

An 8-year retrospective cohort study on 189 Asian patients taking clozapine has found a prevalence of MetSyn of 28.4%; BMI and its changes during the treatment were predictive of MetSyn onset, but hyperglycemia was correlated to treatment duration [148]. Previous studies have shown even higher prevalence values for MetSyn during clozapine treatment: more than 52% [149], 46% in adolescents after 6 months [150], 54% in American outpatients [151], 62% in Australian outpatients [152], and even up to 64% in a small longitudinal, archival, follow-forward study [153].

Regarding specifically type-2 diabetes, a prospective study on 20 patients has shown that clozapine impairs glucose control, but without increasing insulin resistance and independently from BMI increases [154].

Interestingly, it has been proposed that the metabolic side effects of clozapine could be at least in part determined by specific genetic characteristics [17, 18], opening the way for advanced therapy personalization to avoid them. In the last few years, this appears to be the most productive research trend regarding clozapine and MetSyn. However, until now results seem to be still inconclusive [155].

Management of Metabolic Syndrome and Other Metabolic Impairments

Many different solutions have been proposed to the problem of metabolic side effects caused by clozapine. Among them, one can cite combined treatment with fluvoxamine. A 12-week, randomized, double-blind, placebo-controlled study on 85 patients estimated that the combined treatment significantly attenuated the increments in body weight, insulin resistance, and levels of insulin, glucose, and triglycerides compared with clozapine monotherapy [156], without any significant decrease in therapy efficacy. However, more studies are warranted to reach significant conclusions [157].

A meta-analysis has reported that metformin, aripiprazole, and orlistat (in men) are effective therapies [158]. Metformin in particular has been the subject of intense study: As confirmed by another meta-analysis [159], it specifically reduces BMI, waist circumference and weight [160] but has no effects on blood glucose, triglyceride levels, or HDL levels. Some studies have observed some positive effects of metformin on glucose, triglycerides and HDL as well [161, 162].

Therapy with rosiglitazone, topiramate, sibutramine, phenylpropanolamine, modafinil, or atomoxetine produced no significant effect in a meta-analysis, while combined calorie restriction and exercise had limited effects and only in an in-patient setting [158]. At least one study, however, reports some success with rosiglitazone [163].

Comparative Studies

Ziprasidone has a lower tendency to cause MetSyn than clozapine [164]. In a cross-sectional, observational study, clozapine and olanzapine had statistically indistinguishable prevalence of MetSyn (means: 46% for clozapine *vs.* 53% for olanzapine) [165]. In a 6-month, randomized, double-blind, parallel-group trial comparing clozapine to high-dose olanzapine, however, the weight gain was significantly greater for olanzapine [166].

Augmentation Strategies

Augmentation with aripiprazole seems to have limited advantages for MetSyn, in that it can reduce body weight and LDL levels but not glucose and triglyceride levels, and does not increase HDL levels [105]. Moreover, it can exacerbate agitation, akathisia and anxiety [105, 106]. According to another study, aripiprazole can also have beneficial effects on direct cardiovascular parameters such as diastolic blood pressure [107]. Augmentation with aripiprazole while tapering out clozapine to a low dose (50 mg/day) has proven to be useful for metabolic parameters in a single case study [167]. Augmentation with sertindole does not seem to impact metabolic parameters [168]. Augmentation with risperidone could slightly worsen metabolic parameters (mainly blood glucose levels) [169].

In a 1-year study on bipolar disorder, augmenting lithium, valproate or lamotrigine with aripiprazole did not bring about any increase in MetSyn rates, nor in any individual MetSyn parameter [170].

Analytical Methods Suitable for TDM

Several papers on the analysis of clozapine alone or with one or more of its metabolites in human plasma or serum can be found in the literature.

Some older papers report the use of GC [171, 172] methods, sometimes coupled with MS [173]. More recently, several papers have used HPLC-UV [174 - 188] or PDA [189] detection, and other HPLC-ED [190 - 193]. Most methods use acidic mobile phases [180, 182, 184, 185] and a detection wavelength of 230, 240, 254 or 261 nm, while the oxidation potential for amperometric electrochemical detection is 0.7 or 0.8 Volts. The sample pretreatment often consists of LLE [175, 177, 178, 181 - 185, 190, 191] or SPE [181 - 183] procedures. Completely automated analysis with on-line SPE has also been described [174].

A very fast method for the analysis of clozapine and N-desmethylclozapine by means of capillary electrophoresis (CZE) has been developed [194] and seems to be promising for the clinical monitoring of patients.

More recently, methods using LC-MS/MS have become more common [195 - 197] and surely provide outstanding performance, at the cost of more expensive analyses. An automated method using extraction plate technology for sample pre-treatment and flow-injection MS/MS, without HPLC separation, is also available [198]. A novel approach has been developed, based on isotopic internal calibration, for LC-MS/MS methods [199].

Iloperidone

Like other recent atypical antipsychotics, iloperidone (1-[4-[3-[4-(6-fluo-o-1,2-benzoxazol-3-yl)piperidin-1-yl]propoxy]-3-methoxyphenyl]ethenone, Fig. **5**) seems to be associated with a low incidence of weight gain and MetSyn [200]. However, a 25-week, open-label extension study [201] found that weight gain was the second most frequent side effect of the drug (9.2%, after headache), although it was not associated with any significant change in serum glucose, lipid o prolactin levels. The previous, connected double-blind study, which was ziprasidone-controlled [202], reported statistically significant weight gain in 21% of iloperidone-treated patients, as opposed to just 7.4% of those treated with ziprasidone and 3.4 of the placebo controls. However, the results from meta-analyses of clinical trials [203] seem to suggest that weight gain due to iloperidone treatment reaches a plateau after 4-6 weeks of treatment, while the weight gain caused by other antipsychotics (such as olanzapine) seems to be cumulative over time [204]. Weight gain during iloperidone treatment seems also to be dose-related [205] and possibly ethnicity-related [201].

Analytical Methods Suitable for TDM

Some analytical methods for the determination of iloperidone and one [206] or both [207 - 209] of its two main metabolites in plasma have been published. All of them are based on LC-MS/MS, probably due to the very low levels of analytes (less than 10 ng/mL) to be determined in a complex biological matrix. In fact, the limits of quantitation reached by these methods range from 0.01 to 0.25 ng/mL, also aided by the 1:5 sample concentration during pre-treatment.

A paper was published in 1995 on the rapid identification of iloperidone metabolites in blood [210], however it is based on complex hyphenated techniques (such as HPLC - nuclear magnetic resonance (NMR) and LC-MS/MS, or preparative HPLC followed by analytical LC-MS/MS) that seem hardly suitable for extensive application to everyday TDM. Another paper published in 1998 used LC-MS/MS to identify the cytochrome isoforms responsible for iloperidone metabolism [211]; the method would not be immediately applicable to

TDM, but could conceivably be adapted for this purpose without much effort.

Finally, four multi-analyte methods for the simultaneous TDM of several atypical antipsychotics also include iloperidone among the analytes [138, 212 - 214], however it should be noted that two of them [138, 212] only analyze the parent drug, while the other two [213, 214] only include the reduced metabolite P88; while this is surely the most important and active metabolite, dosing both main metabolites would provide a more satisfactory and complete overview of the patients' therapeutic state.

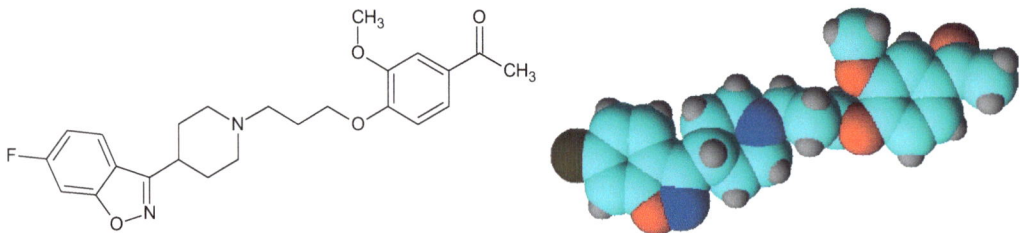

Fig. (5). Conventional chemical structure and 3D-spacefill structure of iloperidone.

Lurasidone

A recent meta-analysis of randomized, controlled trials has concluded that weight gain and BMI during lurasidone ((3a*R*,4*S*,7*R*,7a*S*)-2-((1*R*,2*R*)-2-(4-(1,2-b-nzo-thiazol-3-yl)piperazin-1-ylmethyl)cyclohexylmethyl)hexahyd-o-4,7-methano-2*H*- isoindole-1,3-dione, Fig. **6**) therapy are not significantly different from those of placebo [215]. In the treatment of bipolar disorder, it seems that lurasidone treatment has minimal effects on metabolic parameters and body weight, both in the short [216] and in the long term: A 18-month, open-label continuation study has reported that changes in weight, BMI, lipids, glycemic indices, and prolactin were small, and not clinically meaningful [217]. For psychotic adolescents as well, lurasidone was not associated with significant metabolic changes, as assessed by a 6-week, randomized placebo-controlled study on more than 300 subjects [218]. In acute exacerbation of schizophrenia, lurasidone has confirmed its relatively neutral metabolic profile, although the trial was limited by a high discontinuation rate [219]. Similar results were obtained when lurasidone was used for acute schizophrenia; electrocardiographic parameters were not significantly affected as well [220].

Comparative Studies

A 6-week, double-blind, placebo- and haloperidol-controlled study did not find any significant differences in metabolic parameter changes between lurasidone and either placebo or haloperidol; however, this particular study also failed to separate the therapeutic efficacy of lurasidone and haloperidol from that of placebo. So, it had probably insufficient sensitivity and its results could well be not meaningful for side effects as well [46]. Comparison with extended release quetiapine resulted in minimal changes to all metabolic parameters and BMI in both groups; however, at 6 and 12 months the quetiapine group evidenced significantly higher weight gains [221]. In outpatients with schizophrenia or schizoaffective disorder, neither lurasidone nor ziprasidone produced any clinically significant change in metabolic parameters, except very small reductions in weight and total cholesterol [222].

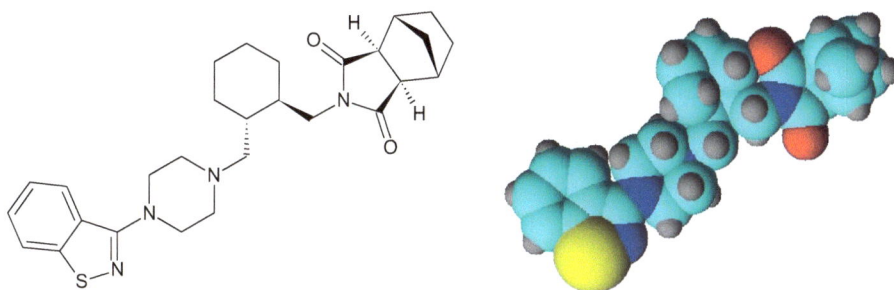

Fig. (6). Conventional chemical structure and 3D-spacefill structure of lurasidone.

Switching Studies

A 6-month, open-label extension study on schizophrenic outpatients who switched from other antipsychotics to lurasidone due to insufficient response did not reveal any clinically relevant adverse changes in body weight, lipids, glucose, insulin, or prolactin [223].

Analytical Methods Suitable for TDM

Three methods possibly suitable for the TDM of lurasidone can be found in the literature [212, 214, 224]. Two of them have been developed by the same research group, and are based on UHPLC-MS/MS. The main difference between the methods is the matrix used: for the first one it is serum, for the second one it is DBS. The third method includes the determination of the main active metabolite, and is thus interesting and advantageous for possible application to TDM. More recently, another LC-MS/MS method has been developed for the determination of both lurasidone and its main active metabolite in human plasma after LLE [224]. Another two methods can be found in the literature for the determination of lurasidone by LC-MS/MS, however they are not applied to human matrices but to rat biological fluids [225, 226]. For this reason, they are not really suitable as such for TDM, but could conceivably be interesting starting points to develop suitable procedures.

Olanzapine

Together with clozapine, olanzapine (2-methyl-4-(4-methylpiperazin-1-yl)-5H-thieno[3,2-c] [1, 5]benzodiazepine, Fig. **7**)) is currently reputed to be one of the antipsychotic drugs most prone to causing MetSyn. Numerous studies are available to support this claim, to the point that only the most recent ones are included in this chapter. As another consequence, recent research is mainly focused on finding the best management tools to prevent or counteract MetSyn (see below), not on studying if and when does it appear during olanzapine treatment.

A recent study on Brazilian patients has shown a large increase in body weight and visceral fat over 12 months of olanzapine treatment, but only a modest increase in blood parameters, such as cholesterol and blood glucose [227]. Thus, the different components of MetSyn seem to be dissociated in their relationship to olanzapine activity. A meta-analysis of randomized studies published between 1992 and 2010 has found that in long-term (more than 48 weeks) studies, the mean weight gain during olanzapine treatment was 5.6 kg. The proportions of patients who gained at least 7, 15 or 25% of their baseline weight with long-term exposure were 64, 32 and 12%, respectively. Elevated indices of glucose metabolism and triglyceride levels have also been observed [228]. In an 8-week study, subjects had significant increases in body weight, triglyceride, total cholesterol, and LDL. Insulin secretion significantly decreased at week 2, returned to baseline at week 4, and significantly increased at week 8; 18.2% of patients met the criteria for significant weight gain and 33.3% met the criteria for MetSyn [229]. These results could point out to a direct activity of olanzapine on pancreatic β-cells.

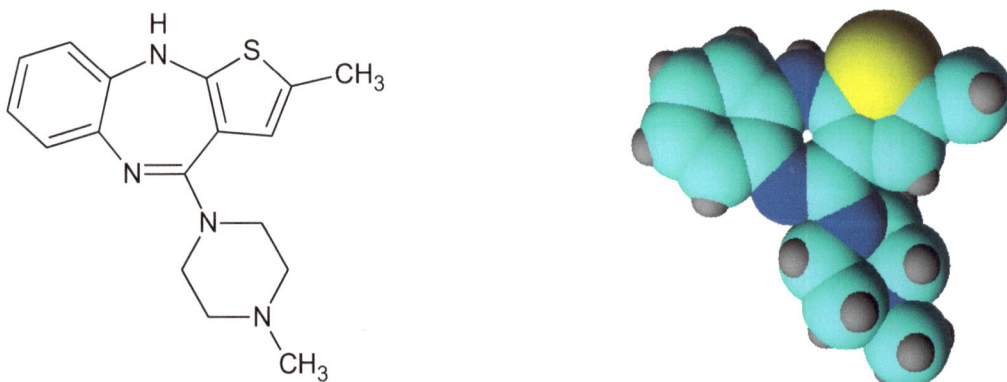

Fig. (7). Conventional chemical structure and 3D-spacefill structure of olanzapine.

A systematic review and meta-analysis of the safety of olanzapine in young children has found 47 studies involving a total of 387 children; weight gain was observed in 78% of patients, and blood glucose elevation in 4% of them [230]. A post-hoc, pooled analysis of 6 randomized, double-blind trials has shown that during olanzapine treatment people of African or Hispanic [231] descent experience weight gain with higher frequency (36%) than Caucasian people (30%) [232].

Interestingly, episodic evidence hints that weight gain, glucose, Hb1A1c, cholesterol and triglyceride level increases can be reverted by olanzapine discontinuation [233].

Management of Metabolic Syndrome and Other Metabolic Impairments

Recently, some authors have suggested that opioidergic transmission may be correlated to metabolic disturbances. For this reasons, different opioid agents have been tested as possible therapeutic agents for olanzapine-induced metabolic side effects. Samidorphan has been tested in a Phase I, randomized, double-blind, placebo-controlled trial. Significantly lower weight gain was reported in the patients treated with samidorphan, however the mean weight gain in these patients was 2.2 kg in 3 weeks [234], so overall efficacy for long-term MetSyn prevention seems to be sub-optimal. Naltrexone has produced significant decreases in fat tissue accumulation and homeostasis model assessment-estimated insulin resistance (HOMA-IR), but no effect on plasma lipid levels [235].

Since low levels of dehydroepiandrosterone are associated with high incidence of MetSyn in the general population, some authors have tested the hypothesis that dehydroepiandrosterone augmentation could be beneficial to olanzapine-treated patients. The study results showed a decrease in glycemia and a stabilization of waist circumference and BMI [236].

Metformin has been also extensively tested. A meta-analysis has included 12 studies, and modest short-term improvements were observed to body weight, waist circumference and BMI [237]. Metformin plus sibutramine has not produced useful effects either, apart for preventing triglyceride increases [238]. However, early trials had obtained more encouraging results, with significant reduction of body weight, glucose, triglyceride, and insulin levels, insulin secretion, and insulin resistance [239]. Better overall results were obtained with topiramate treatment: it resulted in weight loss, decrease in leptin, glucose, cholesterol and triglyceride levels and systolic and diastolic blood pressure [240].

Much interest is currently focused on the possible beneficial activity of natural substances for managing MetSyn during olanzapine treatment. A 12-week, triple-blind, randomized, placebo-controlled study on 66 patients reported positive results from the use of aqueous saffron extract. All patients were without MetSyn at baseline, and at week 12 none of the patients treated with saffron extract developed the illness, as opposed to 9% among those treated with crocin and 27% of those taking placebo. Moreover, both saffron extract and crocin stabilized blood glucose levels [241]. Melatonin has been tested as well in first-episode schizophrenic patients; at week eight, melatonin was associated with significantly less weight gain, increase in waist circumference and triglyceride concentration than placebo. Changes in cholesterol, insulin, and blood sugar concentrations were not significant [242]. Another study on melatonin use in adolescents with bipolar disorder observed lower increases in total cholesterol and systolic blood pressure, but not in triglyceride levels or diastolic blood pressure [243]. On the contrary, omega-3 fatty acids do not seem to produce significant benefits [244].

Comparative Studies

Comparison with the other important MetSyn-causing antipsychotic, clozapine, has often failed to find any difference between these drugs. For example, in a cross-sectional, observational study, clozapine and olanzapine had statistically indistinguishable prevalence of MetSyn [165].

In a 13-week study, both paliperidone and olanzapine caused significant metabolic effects (lipid, glucose and insulin levels were monitored) [245]. A 12-week study has found that weight, BMI, waist and hip circumferences subcutaneous fat, cholesterol, triglycerides, and prolactin all increased in both groups. In contrast, blood glucose, LDL, and homeostasis model assessment-estimated β-cell function (HOMA-B) increased only in the olanzapine group. No differential drug effects over the trial duration were detected on BMI, glucose, HbA1c, insulin, HDL, LDL, cholesterol, triglyceride, or HOMA-IR [246]. A prospective, randomized, controlled trial comparing olanzapine and paliperidone has shown significantly greater undesired metabolic effects during olanzapine treatment [247]. A 2-month comparison between long-acting olanzapine and risperidone did not find any significant differences between the two groups in the prevalence of MetSyn [248], but another study observed significantly higher increases in triglycerides, VLDL and total cholesterol in the olanzapine group [249]. A long-term study observed no differential effects between olanzapine and risperidone on fasting lipid levels after 5 months of treatment. Moreover, there was no difference between treatments on MetSyn development [250].

A 12-week open-label, assessor-blinded randomized trial on 148 patients who were treated with either ziprasidone, olanzapine, olanzapine plus ziprasidone, or switched from olanzapine to ziprasidone, all ziprasidone-treated cohorts experienced less metabolic changes than the olanzapine-treated cohort [251]. In 230 subjects with first-episode schizophrenia during a 6-week, randomized, open-label trial, many MetSyn-related parameters (glucose, insulin, HOMA-IR, LDL, total cholesterol and triglycerides) were significantly lower in the ziprasidone-treated group [252]. On 73 patients with recent-onset schizophrenia or schizoaffective disorder, olanzapine was associated with weight gain and increased triglycerides, cholesterol and transaminases, while ziprasidone was associated with a small decrease in the same parameters. Neither drug affected glucose levels [253].

A large, open-label, randomized study on 300 outpatients with a long history of schizophrenia established that no difference in the metabolic syndrome rate existed between patients treated with haloperidol, olanzapine and aripiprazole [45]. However, a pooled analysis of three randomized clinical trials observed a significantly lower risk of developing MetSyn with aripiprazole [88]. Moreover, both rate and incidence of MetSyn were lower for aripiprazole than for olanzapine in a meta-analysis of four double-blind, randomized, controlled clinical trials [89]. Finally, a classic randomized, double-blind study revealed statistically significant differences in weight change between olanzapine and aripiprazole from week 1 to the endpoint (week 26). Fasting plasma levels of total cholesterol, HDL, and triglycerides were also significantly worse among patients treated with olanzapine [90].

Comparison with quetiapine found lower weight gain, lower increase in calorie intake and lower dyslipidemia for quetiapine [254].

Switching Studies

Switching to asenapine has been successfully used, at least episodically, to revert the weight gain caused by olanzapine therapy [129].

It seems that patients with MetSyn can benefit from switching from olanzapine to aripiprazole: In 62 psychotic patients, those switched to aripiprazole showed continuous improvements in waist circumference, blood pressure, triglyceride levels, blood glucose, and HDL, while those who continued olanzapine treatment experienced a continuous deterioration of the same parameters [95]. In another study, switching to aripiprazole reduced coronary heart disease risk, but without any significant statistical difference in MetSyn prevalence or severity [96].

In an 8-week, prospective, open-label study on bipolar patients, it was observed that switching from olanzapine to ziprasidone produces significant improvements to metabolic parameters [255]. Switching from olanzapine to risperidone can reduce the prevalence of MetSyn (from 54% to 37%) and significantly improve weight, BMI, waist circumference, and systolic and diastolic blood pressure [256].

Overweight patients switching from olanzapine to quetiapine experienced a significant body weight decrease [257].

Augmentation Studies

In an 8-week open-label trial, aripiprazole augmentation brought about significant metabolic advantages for olanzapine-treated patients [103].

Analytical Methods Suitable for TDM

In the past, most methods for the analysis of olanzapine in biological fluids [258 - 274] were based on the use of HPLC - electrochemical detection (ED), due to the drug electroactivity [258 - 261, 264 - 267]; a few HPLC-ED methods have been published in recent years as well [21, 275]. HPLC-UV (or HPLC-PDA) [268 - 271, 273], UHPLC-PDA [271], GC-MS [262, 263, 272] and GC – nitrogen-phosphorus detection (NPD) [274] are well represented too. However, only a few of these methods [21, 265, 266, 275] simultaneously determine olanzapine and its N-desmethyl metabolite. Sample pre-treatment has been carried out mainly by LLE, or by means of SPE procedures using cartridges with different kinds of sorbents [258, 263 - 267, 275]; in the last few years, other, more advanced procedures have been introduced, such as MEPS [272]. An automated HPLC-UV method with column switching pre-treatment is also available [276].

More recently, LC-MS/MS applications have become the norm and a wealth of methods are available in the literature [277 - 284]. Sample pre-treatment is again carried out mainly by LLE [277 - 279, 281 - 283], but papers describing SPE [280] and DPX [284] are also available. However, even using LC-MS, the simultaneous analysis of the parent drug and its main metabolite is carried out by just a fraction of the methods [281].

Quetiapine

According to recent systematic literature reviews, weight gain is one of the most frequent (11-30%) side effects of quetiapine (2-[2-(4-benzo[b] [1, 4]benzothiazepin-6-ylpiperazin-1-yl)ethoxy]ethanol, Fig. **8**) [285]. Episodic reports also corroborate this hypothesis [286, 287]; younger age seems to be correlated to metabolic symptoms and MetSyn onset [288]. Extremely severe episodes are described in the literature, up to fatal acute pancreatitis connected to hypertriglyceridemia and diabetic ketoacidosis [289], especially in case of high quetiapine doses and long-term treatment [290].

In antipsychotic-naïve patients, treatment with quetiapine caused significant increases in body weight and BMI, as well as insulin resistance and secretion. These changes were disconnected from blood glucose, total cholesterol and HDL level changes, which were not significative [291].

Comparative Studies

The common view among researchers is that quetiapine produces metabolic effects that are intermediate between those of clozapine or olanzapine (highest) and those of risperidone (lower) [292 - 294]. On the contrary, some studies, like the 1-year one reported by Emsley *et al.*, have found no weight gain or HgA1c increase during quetiapine treatment [47]. However, this study had a small sample size, so its potency was sub-optimal [295].

The double-blind, randomized TEA study comparing quetiapine ER to aripiprazole in 113 children and adolescents with first-episode psychosis concluded that quetiapine is associated with more metabolic side effects, while aripiprazole is associated with more initial akathisia and sedation [87]. A 2010 review of the studies included in the Cochrane Schizophrenia Group Trials Register comparing quetiapine with other antipsychotics showed that quetiapine may cause less weight gain than olanzapine and paliperidone, more cholesterol increase than risperidone, and more weight gain and cholesterol increase than ziprasidone. Moreover, it appears to have a similar weight gain profile to risperidone, as well as clozapine and aripiprazole (but the latter two comparisons are based on very limited data) [296]. Comparison with olanzapine found lower weight gain, lower increase in calorie intake and lower dyslipidemia for quetiapine [254].

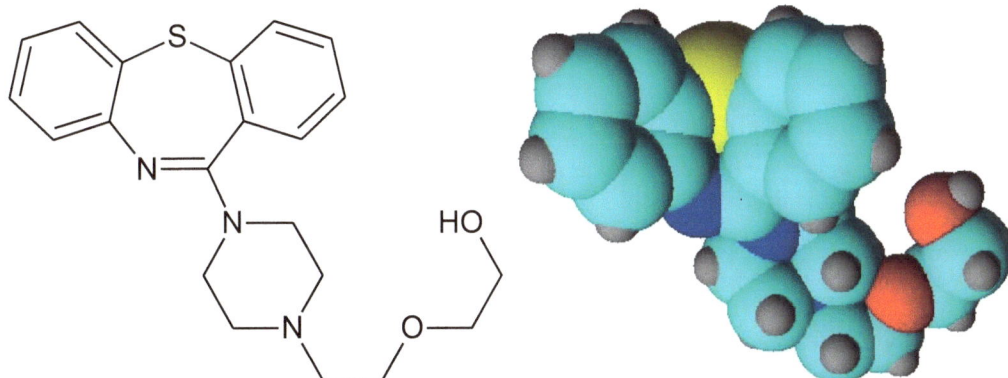

Fig. (8). Conventional chemical structure and 3D-spacefill structure of quetiapine.

Switching Studies

Switching from quetiapine to aripiprazole seems to reduce coronary heart disease risk, but without any significant statistical difference in MetSyn prevalence or severity [96]. Patients switching from quetiapine to ziprasidone evidenced small but significant decreases in body weight as well as improved lipid profiles [297]. Overweight patients switching from olanzapine to quetiapine also experienced a body weight decrease (mean of 2.3 kg after 10 weeks) [257].

Augmentation Strategies

Augmentation of quetiapine treatment with valproate can lead to severe hypertriglyceridemia and hypercholesterolemia [298].

Analytical Methods Suitable for TDM

An HPLC method has been published for pharmacokinetic studies of quetiapine and its two metabolites 7-hydroxyquetiapine and 7-hydroxy-*N*-dealkylquetiapine in plasma [299]. More recently, other selective and sensitive HPLC methods for the analysis of quetiapine in plasma of patients have been published: using HPLC-UV after LLE [300], SPE [301, 302] or thin-film microextraction (TFME) [303] pre-treatment. HPLC-UV coupled to column-switching for sample purification has been described as well [276].

As usual, in the last few years LC-MS/MS is becoming the golden standard for the analysis of quetiapine in patient plasma, using ESI-TQ after LLE [304] or SPE [305, 306], after PPP in a 96-well plate [307], or DPX with mixed-mode restricted access material (RAM) and C18 particles [284]. LC-MS/MS was also used for the analysis of quetiapine together with its metabolite norquetiapine [308], or with four metabolites [309].

Risperidone

Risperidone (3-[2- [4-(6-fluoro-1, 2-benzoxazol-3-yl) piperidin-1-yl]ethyl]-2-methyl-6,7,8,9-tetrahydropyrido[1,2-a]pyrimidin-4-one, Fig. (**9**)) seems to have an especially strong effect on young people's metabolism. In 97 children with autism spectrum disorder, risperidone has shown a tendency to cause increases to BMI, levels of glucose, HbA1c, insulin; after 16 weeks, the number of children with MetSyn increased by 71% (from 7 to 12) [310]. Other studies confirm this finding: a 30-week, prospective, observational study on 22 children and adolescents has found that risperidone treatment can induce statistically significant increases in waist circumference, body weight, BMI and total cholesterol, as well as changes in the distribution of the abdominal fat mass and constitution of the liver parenchyma [311]; another study on 26 children and adolescents confirmed a strong link between risperidone treatment and obesity risk [312]. Other authors have found that risperidone treatment is associated with a number of metabolic abnormalities, especially when the patient's weight is already above normal at the beginning of the therapy; however, the prevalence of MetSyn is low [313].

Interestingly, these changes seem to be reversible with therapy discontinuation [314]: a long-term study has found that after several years of treatment with risperidone, discontinuing the antipsychotic therapy produces a reversal of weight gain, and a corresponding improvement in cardiometabolic parameters. However, this does not generally happen in patients who switch to other atypical antipsychotics [315].

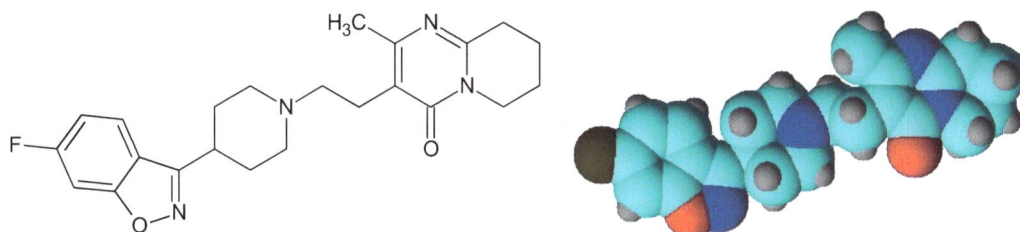

Fig. (9). Conventional chemical structure and 3D-spacefill structure of risperidone.

Comparative Studies

A recent study on patients in therapy with long-acting injectable olanzapine or risperidone found no difference in MetSyn induction between the two drugs, but a significantly higher frequency of cardiovascular risk factors (hypertension, contractility anomalies, diastolic dysfunction) in risperidone-treated patients [248]. In a large-scale 26-week, randomized, double-blind, controlled comparative trial of 457 patients taking either cariprazine or risperidone, no clear difference in metabolic parameter changes (total cholesterol, LDL, HDL, triglycerides, glucose, blood pressure, body weight, BMI, waist circumference) was seen between the two drugs; no significant change in comparison to the baseline was observed either [316]. Another study on 60 patients has found that olanzapine has a higher propensity to alter lipid profile parameters than risperidone (especially triglycerides, VLDL and total cholesterol) [249]. A long-term study observed that there were no differential drug effects on fasting lipid levels after 5 months of treatment. Moreover, there was no difference between treatments on development of MetSyn during the 5-month period [250]. In child and adolescent inpatients, treatment with both olanzapine and risperidone is correlated with a significant increase in BMI; however, risperidone does not increase risk factors for diabetes and metabolic syndrome as olanzapine does [317].

When comparing risperidone with paliperidone (which is also the main risperidone metabolite, 9-hydroxyrisperidone), no significant differences in mean weight change, most metabolic parameters, or mean efficacy measures were observed at end point in several different studies [318, 319]. A study that compared aripiprazole, olanzapine and risperidone obtained a significantly lower risk of developing MetSyn for aripiprazole-treated patients. However, after adjusting the results for age and gender, no difference was found among the three antipsychotics [93]. In a short- and long-term comparison study with risperidone, sertindole caused similar body weight and BMI increases but, like risperidone, did not increase the risk of MetSyn [320].

Comparison between risperidone and ziprasidone proved to be advantageous for ziprasidone regarding weight gain in a long-term 44-week, randomized, double-blind continuation trial on 139 subjects [321].

Switching Studies

Switching from risperidone to aripiprazole seems to reduce coronary heart disease risk, but without any significant statistical difference in MetSyn prevalence or severity [96]. Switching from risperidone to ziprasidone can result in weight loss, and this has been observed in more than one study [49, 322]. Switching from olanzapine to risperidone can reduce the prevalence of MetSyn (from 54% to 37%) and significantly improve weight, BMI, waist circumference, and systolic and diastolic blood pressure [256].

Augmentation Strategies

Augmenting risperidone with aripiprazole does not lead to any significant change in plasma glucose, total cholesterol, triglyceride and HDL levels, and modestly reduces prolactin levels, but significantly decreases serum LDL and weight gain [104]. Comparison of risperidone with low-dose risperidone augmented with low-dose haloperidol showed no difference in changes in weight, vital signs, corrected QT interval, liver/renal function, fasting glucose level, and lipid profiles; risperidone alone was associated with higher levels of prolactin [323].

Analytical Methods Suitable for TDM

Since the "active moiety" during risperidone therapy is the sum of risperidone and its 9-hydroxy metabolite, almost all analytical methods developed for the therapeutic drug monitoring of this drug simultaneously determine the two active species. Most of these methods use chromatography coupled with either UV [324 - 326] or coulometric ED [328 - 334] detection, in reversed- or direct-phase mode, with mobile phases having different acid/base properties. More recently, UHPLC-

PDA has also been introduced for very fast analysis [271]. The detection wavelengths of the UV methods are 278 nm [324] or 280 nm [325, 326] while the voltages applied to the coulometric detector are in the 0.50 - 0.60 Volt range for the first analytical cell and in the 0.80 - 0.96 Volt range for the second analytical cell. Recently, LC-MS/MS is becoming the most common technique in the literature: *e.g.*, ESI-MS/MS for the analysis of blood [335] or plasma [336 - 338], ESI-TQ for analysis of hair [339], ESI-TQ for plasma and saliva [340], ESI-QTRAP [341] or APCI-TQ [342] for plasma.

Most methods use sample pre-treatment procedures based on liquid-liquid extraction [324, 326, 328, 330, 335, 337, 341, 342] which may be followed by one or more back extractions. Only a few papers [325, 327, 334] use SPE procedures for the pre-treatment of patient plasma, or MEPS for pre-treatment miniaturization [332]. LC-MS/MS methods, due to their intrinsic selectivity and sensitivity, often use simplified pre-treatment procedures such as PPP [336, 338, 341].

Since the 9-hydroxy metabolite is a chiral molecule, some studies also performed the determination of its enantiomers (as well as risperidone) in patient plasma by HPLC-ED [343] or LC-MS/MS [344, 345].

Sertindole

Among atypical antipsychotics, sertindole (1-[2-[4-[5-chloro-1-(4-fluorophe-nyl) indol-3-yl]piperidin-1-yl]ethyl]imidazolidin-2-one, Fig. **10**) is in a peculiar position, since it is known to cause QTc interval prolongation and possibly arrhythmias as a consequence. For this reason, it has not been approved in the USA yet, and is available only in single countries in Europe, and only associated with a strict program of cardiac function monitoring [346].

Regarding MetSyn specifically, however, sertindole does not seem to be less safe than other atypical antipsychotics [347]. It can cause modest increases in blood glucose [348], triglyceride [348] and cholesterol [349, 350] levels, and also significant body weight gain [351], which was observed also in comparison with haloperidol [352]. In a short- and long-term comparison study with risperidone, sertindole caused similar body weight and BMI increases but, like risperidone, did not increase the risk of MetSyn [320].

A 12-week, double-blind, randomized, placebo-controlled study on augmentation of clozapine therapy with sertindole did not find any significant benefit on schizophrenia symptoms; however, no metabolic parameter was changed by the introduction of sertindole [168].

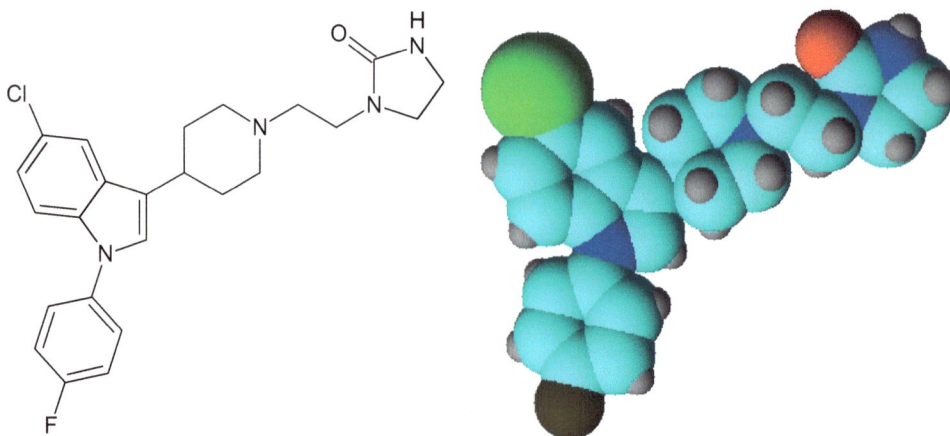

Fig. (10). Conventional chemical structure and 3D-spacefill structure of sertindole.

Analytical Methods Suitable for TDM

Due to the limited diffusion of this drug, and its unavailability in the USA, analytical methods are few and far between. A method is based on HPLC-FL after column-switching purification [353]. Some papers report the simultaneous analysis of the parent drug together with its main metabolites, dehydrosertindole and norsertindole, by HPLC-UV after LLE [354], or by LC-MS/MS [355].

A voltammetric method using glassy carbon (for differential pulse) and boron-doped diamond (for square wave) electrodes has been published as well [356]. Sample pre-treatment is carried out by PPP with acetonitrile.

Ziprasidone

Ziprasidone (5-[2-[4-(1,2-benzothiazol-3-yl)piperazin-1-yl]ethyl]-6-chloro-1,3 - di-hydroindol-2-one, Fig. (**11**)) was found to have a neutral weight and metabolic profile in adolescents during a 6-week, randomized, double-blind, placebo-controlled multicenter study followed by a 26-week open-label extension trial, although this study failed to find any efficacy (better than placebo) in treating schizophrenia with ziprasidone [357].

Ziprasidone has been tested as an adjunctive therapy to lithium or valproate in the treatment of bipolar disorder. In a study on 584 patients, the ziprasidone plus mood stabilizer group was not distinguishable from the placebo plus mood stabilizer group in relation to body weight and metabolic profiles [358].

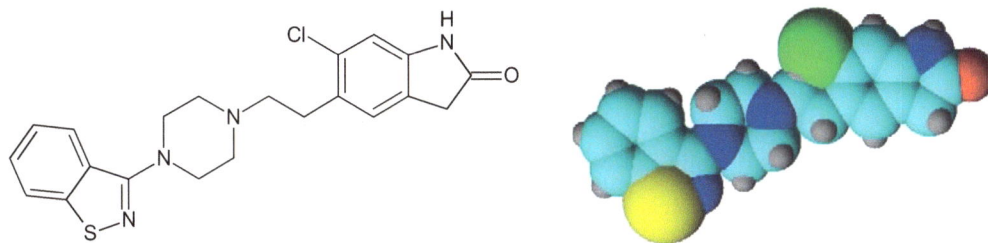

Fig. (11). Conventional chemical structure and 3D-spacefill structure of ziprasidone.

Comparative Studies

A 52-week, follow-up, comparative study on the efficacy and tolerability of paliperidone (extended release), aripiprazole and ziprasidone on 203 patients with first-episode schizophrenia has found that efficacy and metabolic disruption are approximately inversely proportional, with paliperidone having the highest efficacy and the highest propensity to impair lipid metabolism (but not glucose metabolism), and aripiprazole having negative effects on glucose metabolism (but not on lipid metabolism). Ziprasidone had the lowest efficacy and also the lowest propensity to influence lipids and glucose levels [92]. A comparison of lurasidone and ziprasidone did not find any significant differences in safety between them. Both drugs caused small reductions in body weight and total cholesterol, while no change was observed in triglyceride levels [222]. Comparison with risperidone proved to be advantageous for ziprasidone regarding weight gain in a long-term 44-week, randomized, double-blind continuation trial on 139 subjects [321].

Large-scale studies have confirmed that ziprasidone is relatively safe in relation to MetSyn [359]. For example, the CATIE study on almost 1500 patients (comparing olanzapine, quetiapine, risperidone, ziprasidone and perphenazine) showed that ziprasidone decreased Hb1A1c, total cholesterol and triglycerides, also significantly decreasing the prevalence of MetSyn in comparison to olanzapine [360]. An epidemiological study on more than 55 000 schizophrenic patients found no increased risk of diabetes from ziprasidone treatment, while the risk was increased by olanzapine and clozapine [361]. The 12-months EUFEST study on almost 500 patients treated with haloperidol, amisulpride, olanzapine, quetiapine or ziprasidone showed no increased metabolic risk factor for ziprasidone, with the drug being amongst those with the lowest propensity to cause weight gain [362]. A 6-week, multicenter, double-blind, parallel-design, flexible-dose trial on more than 160 patients revealed that olanzapine and ziprasidone had similar primary and secondary efficacy endpoints, but with significant advantages in the ziprasidone arm regarding body weight, total cholesterol, triglycerides, LDL and fasting insulin levels [363]. However, another study on schizophrenic patients with prominent depressive symptoms observed similar effects on metabolism (olanzapine caused greater increases in triglycerides, glycosylated hemoglobin, and weight) but different efficacy measurements: Over 24 weeks, olanzapine caused significantly greater improvements in depressive symptoms, higher proportions of patients completing the study and longer permanence of patients on medication [364]. Olanzapine was directly compared to ziprasidone during a 6-week, randomized, open-label trial on 230 subjects with first-episode schizophrenia. Many MetSyn-related parameters were significantly lower in the ziprasidone-treated group: fasting plasma glucose, insulin, homeostasis model assessment 2-insulin resistance, LDL, total cholesterol and triglycerides [252], thus highlighting the better metabolic profile of ziprasidone. Similar results were obtained on 73 patients with recent-onset schizophrenia or schizoaffective disorder in an 8-week double-blind randomized controlled trial: olanzapine was associated with weight gain and increased triglycerides, cholesterol and transaminases, while ziprasidone was associated with a small decrease in the same parameters. Neither drug affected glucose levels [253].

In any case, episodic reports of patients developing MetSyn after switching from other antipsychotics to ziprasidone [365] also exist. An observational, non-randomized study [366] found a higher incidence of diabetes in patients treated with ziprasidone than in those treated with aripiprazole and other second-generation antipsychotics. However, the study was not designed to test the specific causality between a single antipsychotic and diabetes, so this result can only be used to understand how clinicians treat their schizophrenic patients and how this affects them, as also observed by Vanderburg *et al.* [359].

Switching Studies

Since ziprasidone does not seem to cause strong metabolic effects [367], it has been proposed as a possible second choice in switching from olanzapine, or as a combined agent. A 12-week open-label, assessor-blinded randomized trial on 148 patients who were treated with either ziprasidone alone, olanzapine alone, low-dose olanzapine plus low-dose ziprasidone, or switched from olanzapine to ziprasidone, all ziprasidone-treated cohorts experienced less metabolic changes than the olanzapine- treated cohort. In particular, weight gain, BMI increase, blood glucose increase and total blood cholesterol increase were all significantly lower in the three ziprasidone-treated cohorts [251]. Best efficacy/tolerability combination was obtained in the olanzapine plus ziprasidone cohort. In another 8-week, prospective, open-label study on bipolar patients, it was similarly observed that switching from olanzapine to ziprasidone brings about significant improvements to metabolic parameters [255].

A 1-year, open-label, prospective study on 114 patients who switched from other antipsychotics to ziprasidone found that about half of them (59%) had one less risk factor for MetSyn at the endpoint; moreover, the patients' 10-year coronary heart disease risk decreased continually during the study. Reductions in other metabolic parameters (blood LDL, total cholesterol, insulin, glycated hemoglobin, body weight and BMI) were also observed [368]. Similarly, significant improvements were found for total cholesterol, low-density lipoprotein, high-density lipoprotein, triglycerides and body weight when switching from typical and/or atypical antipsychotics to ziprasidone in an 8-week, open-label, flexible dose study on 312 patients who were non-responders or intolerant to treatment [369]. Significant baseline to endpoint reductions were seen in mean weight, BMI, glucose, total cholesterol, and triglyceride levels after switching to ziprasidone in a 6-month study on 84 patients [370].

A 63-day, open-label, prospective study was on 26 inpatients with type II diabetes switching from antipsychotic polypharmacy to ziprasidone monotherapy reported notable reductions in fasting glucose, capillary blood glucose, weight and BMI. There was also a reduction in the use of antidiabetic medications and in the number of patients meeting criteria for MetSyn (from 9 to 4). However, 10 of the 26 patients failed to complete the study due to psychotic relapse [371]. Switching from quetiapine to ziprasidone produced small but significant decreases in total cholesterol, LDL and triglycerides, but not in blood glucose or glycated hemoglobin after 16 weeks [297] in 241 subjects. Switching from aripiprazole to ziprasidone significantly decreased body weight, waist and hip circumferences, fasting blood glucose, and alanine aminotransferase in a 12-week, open-label study on 19 patients, despite the fact that aripiprazole is generally acknowledged as an antipsychotic relatively devoid of metabolic side effects [101]. Significant reductions in weight, fasting glucose, total cholesterol, LDL, and triglycerides were found at the endpoint when switching from clozapine to ziprasidone in an 18-week, randomized, flexible-dose, double-blind, double-dummy trial on 147 treatment-refractory schizophrenic patients (the MOZART study) [164]. Previously, it had already observed that switching from olanzapine or risperidone to ziprasidone was associated with clinically significant improvements in weight, BMI, total cholesterol, and triglyceride levels, and that these improvements were sustained for 52 weeks. On the contrary, switching from typical antipsychotics was not associated with significant changes in weight or lipid parameters [372].

Analytical Methods Suitable for TDM

The published methods for the analysis of ziprasidone in biological fluids are mainly based on HPLC-UV [373 - 375], LC-MS/MS [376 - 378], or UHPLC-MS/MS [379]; a few exploit specific electrodes and potentiometry [380, 381].

Two papers also describe HPLC-FL methods for ziprasidone determination in plasma [382, 383]. The last one proposes a fast procedure for microsampling and purification based on DBS and solvent extraction [383]. Most of the other methods include a variety of pre-treatment procedures, ranging from PPP [379], LLE [374] or SPE [373], to SPME [375] and column switching [373].

Some multianalyte methods also include ziprasidone; for example, a UHPLC-MS/MS method coupled to on-line SPE pre-treatment [384].

CONCLUDING REMARKS

MetSyn is currently among the most worrisome side effects of atypical antipsychotic treatment and one of the main causes of patient non-compliance and treatment discontinuation. Although no scientific consensus has been reached yet, it seems that some of the most recent atypical antipsychotics (aripiprazole, asenapine, ziprasidone) can have a reduced incidence of MetSyn when compared to older ones (clozapine, olanzapine, risperidone). The former can thus be used as the first therapeutic choice, or as possible alternatives for switching or augmentation in case of therapy intolerance to the latter. However, more, larger studies are needed to provide more solid, useful guidelines for clinicians. One of the tools most suitable for this purpose is TDM. If it is carried out routinely for most psychiatric patients undergoing pharmacotherapy, TDM can provide large amounts of very reliable information on CCC, leading to more informed decisions on the clinician's part and to considerably less attrition in therapeutic outcomes. Many different analytical methods are available for this purpose, using instrumentation that can be found in most centralized bioanalytical laboratories, and also including very advantageous microsampling and automation options.

CONSENT FOR PUBLICATION

Not applicable.

CONFLICT OF INTEREST

The authors confirm that they have no conflict of interest to declare for this publication.

ACKNOWLEDGEMENTS

Declared none.

REFERENCES

[1]   Steel Z, Marnane C, Iranpour C, *et al.* The global prevalence of common mental disorders: a systematic review and meta-analysis 1980-2013. Int J Epidemiol 2014; 43(2): 476-93.

[http://dx.doi.org/10.1093/ije/dyu038] [PMID: 24648481]

[2]   Lieberman JA, Stroup TS, McEvoy JP, *et al.* Effectiveness of antipsychotic drugs in patients with chronic schizophrenia. N Engl J Med 2005; 353(12): 1209-23.
[http://dx.doi.org/10.1056/NEJMoa051688] [PMID: 16172203]

[3]   Vita A, Barlati S, Deste G, Corsini P, De Peri L, Sacchetti E. Factors related to different reasons for antipsychotic drug discontinuation in the treatment of schizophrenia: a naturalistic 18-month follow-up study. Psychiatry Res 2012; 200(2-3): 96-101.
[http://dx.doi.org/10.1016/j.psychres.2012.07.006] [PMID: 22858250]

[4]   Yang S-Y, Chen L-Y, Najoan E, *et al.* Polypharmacy and psychotropic drug loading in patients with schizophrenia in Asian countries: Fourth survey of Research on Asian Prescription Patterns on antipsychotics. Psychiatry Clin Neurosci 2018; 72(8): 572-9.
[http://dx.doi.org/10.1111/pcn.12676] [PMID: 29761577]

[5]   Tiihonen J, Tanskanen A, Taipale H. 20-Year Nationwide Follow-Up Study on Discontinuation of Antipsychotic Treatment in First-Episode Schizophrenia. Am J Psychiatry 2018; 175(8): 765-73.
[http://dx.doi.org/10.1176/appi.ajp.2018.17091001] [PMID: 29621900]

[6]   Bowtell M, Eaton S, Thien K, *et al.* Rates and predictors of relapse following discontinuation of antipsychotic medication after a first episode of psychosis. Schizophr Res 2018; 195: 231-6.
[http://dx.doi.org/10.1016/j.schres.2017.10.030] [PMID: 29066258]

[7]   Han C, Rice MW, Cai D. Neuroinflammatory and autonomic mechanisms in diabetes and hypertension. Am J Physiol Endocrinol Metab 2016; 311(1): E32-41.
[http://dx.doi.org/10.1152/ajpendo.00012.2016] [PMID: 27166279]

[8]   Guarino D, Nannipieri M, Iervasi G, Taddei S, Bruno RM. The role of the autonomic nervous system in the pathophysiology of obesity. Front Physiol 2017; 8: 665.
[http://dx.doi.org/10.3389/fphys.2017.00665] [PMID: 28966594]

[9]   Definition, diagnosis and classification of diabetes mellitus and its complications: Part 1: diagnosis and classification of diabetes mellitus. Geneva, Switzerland: Department of Noncommunicable Disease Surveillance 1999.

[10]   Expert Panel on Detection, Evaluation, and Treatment of High Blood Cholesterol in Adults. JAMA 2001; 285: 2486-97.
[http://dx.doi.org/10.1001/jama.285.19.2486] [PMID: 11368702]

[11]   Alberti KGMM, Zimmet P, Shaw J. Metabolic syndrome--a new world-wide definition. A Consensus Statement from the International Diabetes Federation. Diabet Med 2006; 23(5): 469-80.
[http://dx.doi.org/10.1111/j.1464-5491.2006.01858.x] [PMID: 16681555]

[12]   Hu G, Qiao Q, Tuomilehto J, Balkau B, Borch-Johnsen K, Pyorala K. Prevalence of the metabolic syndrome and its relation to all-cause and cardiovascular mortality in nondiabetic European men and women. Arch Intern Med 2004; 164(10): 1066-76.
[http://dx.doi.org/10.1001/archinte.164.10.1066] [PMID: 15159263]

[13]   Cameron AJ, Shaw JE, Zimmet PZ. The metabolic syndrome: prevalence in worldwide populations. Endocrinol Metab Clin North Am 2004; 33(2): 351-75.
[http://dx.doi.org/10.1016/j.ecl.2004.03.005] [PMID: 15158523]

[14]   Ford ES, Giles WH, Dietz WH. Prevalence of the metabolic syndrome among US adults: findings from the third National Health and Nutrition Examination Survey. JAMA 2002; 287(3): 356-9.
[http://dx.doi.org/10.1001/jama.287.3.356] [PMID: 11790215]

[15]   Amzand SG, Luteijn BL, van der Ven E, Bogers JP, Selten J-P. Diagnostic value of a simplified screening test for metabolic syndrome in a Dutch patient cohort with schizophrenia spectrum disorders. Australas Psychiatry 2018; 26(6): 615-8.
[http://dx.doi.org/10.1177/1039856218779137] [PMID: 29888619]

[16]   Lee AMH, Ng CG, Koh OH, Gill JS, Aziz SA. Metabolic syndrome in first episode schizophrenia,

based on the national mental health registry of schizophrenia (NMHR) in a general hospital in Malaysia: A 10-year retrospective cohort study. Int J Environ Res Public Health 2018; 15(5): 933.
[http://dx.doi.org/10.3390/ijerph15050933] [PMID: 29735938]

[17]   Vasudev K, Choi Y-H, Norman R, Kim RB, Schwarz UI. Genetic Determinants of Clozapine-Induced Metabolic Side Effects. Can J Psychiatry 2017; 62(2): 138-49.
[http://dx.doi.org/10.1177/0706743716670128] [PMID: 27681143]

[18]   Piatkov I, Caetano D, Assur Y, *et al.* CYP2C19*17 protects against metabolic complications of clozapine treatment. World J Biol Psychiatry 2017; 18(7): 521-7.
[http://dx.doi.org/10.1080/15622975.2017.1347712] [PMID: 28664816]

[19]   Dos Santos-Júnior A, Tamascia ML, Lorenzetti R, *et al.* Serum Concentration of Risperidone and Adverse Effects in Children and Adolescents. J Child Adolesc Psychopharmacol 2017; 27(2): 211-2.
[http://dx.doi.org/10.1089/cap.2016.0114] [PMID: 27779425]

[20]   Honer WG, MacEwan GW, Gendron A, *et al.* A randomized, double-blind, placebo-controlled study of the safety and tolerability of high-dose quetiapine in patients with persistent symptoms of schizophrenia or schizoaffective disorder. J Clin Psychiatry 2012; 73(1): 13-20.
[http://dx.doi.org/10.4088/JCP.10m06194] [PMID: 21733490]

[21]   Lu M-L, Lin C-H, Chen Y-C, Yang H-C, Wu T-H. Determination of olanzapine and N-desmethy--olanzapine in plasma using a reversed-phase HPLC coupled with coulochemical detection: correlation of olanzapine or N-desmethyl-olanzapine concentration with metabolic parameters. PLoS One 2013; 8(5)e65719
[http://dx.doi.org/10.1371/journal.pone.0065719] [PMID: 23741510]

[22]   Lu M-L, Chen C-H, Kuo P-T, Lin C-H, Wu T-H. Application of plasma levels of olanzapine and N-desmethyl-olanzapine to monitor metabolic parameters in patients with schizophrenia. Schizophr Res 2018; 193: 139-45.
[http://dx.doi.org/10.1016/j.schres.2017.07.022] [PMID: 28720417]

[23]   Hiemke C, Baumann P, Bergemann N, *et al.* AGNP Consensus Guidelines for Therapeutic Drug Monitoring in Psychiatry: Update 2011. Pharmacopsychiatry 2011; 44: 195-235.
[http://dx.doi.org/10.1055/s-0031-1286287] [PMID: 22053351]

[24]   Best-Shaw L, Gudbrandsen M, Nagar J, Rose D, David AS, Patel MX. Psychiatrists' perspectives on antipsychotic dose and the role of plasma concentration therapeutic drug monitoring. Ther Drug Monit 2014; 36(4): 486-93.
[http://dx.doi.org/10.1097/FTD.0000000000000041] [PMID: 24384695]

[25]   Mandrioli R, Protti M, Mercolini L. Novel atypical antipsychotics: Metabolism and therapeutic drug monitoring (TDM). Curr Drug Metab 2015; 16(2): 141-51.
[http://dx.doi.org/10.2174/1389200216666150602145005] [PMID: 26033329]

[26]   Raggi MA, Mandrioli R, Sabbioni C, Pucci V. Atypical antipsychotics: pharmacokinetics, therapeutic drug monitoring and pharmacological interactions. Curr Med Chem 2004; 11(3): 279-96.
[http://dx.doi.org/10.2174/0929867043456089] [PMID: 14965232]

[27]   Mandrioli R, Mercolini L, Saracino MA, Raggi MA. Selective serotonin reuptake inhibitors (SSRIs): therapeutic drug monitoring and pharmacological interactions. Curr Med Chem 2012; 19(12): 1846-63.
[http://dx.doi.org/10.2174/092986712800099749] [PMID: 22414078]

[28]   Müller MJ, Dragicevic A, Fric M, *et al.* Therapeutic drug monitoring of tricyclic antidepressants: how does it work under clinical conditions? Pharmacopsychiatry 2003; 36(3): 98-104.
[http://dx.doi.org/10.1055/s-2003-39983] [PMID: 12806567]

[29]   Mercolini L, Mandrioli R, Finizio G, Boncompagni G, Raggi MA. Simultaneous HPLC determination of 14 tricyclic antidepressants and metabolites in human plasma. J Sep Sci 2010; 33(1): 23-30.
[http://dx.doi.org/10.1002/jssc.200900493] [PMID: 20091716]

[30]   Mandrioli R, Protti M, Mercolini L. New-Generation, Non-SSRI Antidepressants: Therapeutic Drug Monitoring and Pharmacological Interactions. Part 1: SNRIs, SMSs, SARIs. Curr Med Chem 2018; 25(7): 772-92.
[http://dx.doi.org/10.2174/0929867324666170712165042] [PMID: 28707591]

[31]   Mandrioli R, Mercolini L. Metabolism of drugs used in the therapy of seizures: an analytical point of view. Part 1. Curr Drug Metab 2017; 18(8): 735-56.
[http://dx.doi.org/10.2174/1389200218666170406115350] [PMID: 28382861]

[32]   Perry PJ. Therapeutic drug monitoring of antipsychotics. Psychopharmacol Bull 2001; 35(3): 19-29.
[PMID: 12397876]

[33]   Mandrioli R, Mercolini L, Raggi MA. Benzodiazepine metabolism: an analytical perspective. Curr Drug Metab 2008; 9(8): 827-44.
[http://dx.doi.org/10.2174/138920008786049258] [PMID: 18855614]

[34]   Mercolini L, Mandrioli R, Amore M, Raggi MA. Separation and HPLC analysis of 15 benzodiazepines in human plasma. J Sep Sci 2008; 31(14): 2619-26.
[http://dx.doi.org/10.1002/jssc.200800212] [PMID: 18693306]

[35]   Mandrioli R, Mercolini L, Raggi MA. Metabolism of benzodiazepine and non-benzodiazepine anxiolytic-hypnotic drugs: an analytical point of view. Curr Drug Metab 2010; 11(9): 815-29.
[http://dx.doi.org/10.2174/138920010794328887] [PMID: 21189133]

[36]   Grundmann M, Kacirova I, Urinovska R. Therapeutic drug monitoring of atypical antipsychotic drugs. Acta Pharm 2014; 64(4): 387-401.
[http://dx.doi.org/10.2478/acph-2014-0036] [PMID: 25531781]

[37]   Hiemke C, Dragicevic A, Gründer G, *et al.* Therapeutic monitoring of new antipsychotic drugs. Ther Drug Monit 2004; 26(2): 156-60.
[http://dx.doi.org/10.1097/00007691-200404000-00012] [PMID: 15228157]

[38]   Patteet L, Maudens KE, Stove CP, *et al.* Are capillary DBS applicable for therapeutic drug monitoring of common antipsychotics? A proof of concept. Bioanalysis 2015; 7(16): 2119-30.
[http://dx.doi.org/10.4155/bio.15.100] [PMID: 26327190]

[39]   Fragou D, Dotsika S, Sarafidou P, Samanidou V, Njau S, Kovatsi L. Atypical antipsychotics: trends in analysis and sample preparation of various biological samples. Bioanalysis 2012; 4(8): 961-80.
[http://dx.doi.org/10.4155/bio.12.55] [PMID: 22533569]

[40]   Mercolini L, Saracino MA, Protti M. Current advances in biosampling for therapeutic drug monitoring of psychiatric CNS drugs. Bioanalysis 2015; 7(15): 1925-42.
[http://dx.doi.org/10.4155/bio.15.123] [PMID: 26295991]

[41]   Klett CJ, Caffey EM Jr. Weight changes during treatment with phenothiazine derivatives. J Neuropsychiatry 1960; 2: 102-8.
[PMID: 13756782]

[42]   Amdisen A. Drug-produced obesity. Experiences with Chlorpromazine, Perphenazine and Clopenthixol. Dan Med Bull 1964; 11: 182-9.
[PMID: 14209068]

[43]   Cao D, Xie S-P, Chen Q-B, Yuan Y-G, Fang Q. Characteristics of the sexual disturbance caused by chlorpromazine, risperidone, quetiapine and olanzapine and their associations with the changes of blood glucose and blood lipids in male patients with schizophrenia. Zhongguo Linchuang Kangfu 2005; 9(36): 63-8.

[44]   Amamoto T, Kumai T, Nakaya S, Matsumoto N, Tsuzuki Y, Kobayashi S. The elucidation of the mechanism of weight gain and glucose tolerance abnormalities induced by chlorpromazine. J Pharmacol Sci 2006; 102(2): 213-9.
[http://dx.doi.org/10.1254/jphs.FP0060673] [PMID: 17031068]

[45] Parabiaghi A, Tettamanti M, D'Avanzo B, *et al.* Metabolic syndrome and drug discontinuation in schizophrenia: a randomized trial comparing aripiprazole olanzapine and haloperidol. Acta Psychiatr Scand 2016; 133(1): 63-75.
[http://dx.doi.org/10.1111/acps.12468] [PMID: 26252780]

[46] Potkin SG, Kimura T, Guarino J. A 6-week, double-blind, placebo- and haloperidol-controlled, phase II study of lurasidone in patients with acute schizophrenia. Ther Adv Psychopharmacol 2015; 5(6): 322-31.
[http://dx.doi.org/10.1177/2045125315606027] [PMID: 26834965]

[47] Emsley R, Turner HJ, Schronen J, Botha K, Smit R, Oosthuizen PP. Effects of quetiapine and haloperidol on body mass index and glycaemic control: a long-term, randomized, controlled trial. Int J Neuropsychopharmacol 2005; 8(2): 175-82.
[http://dx.doi.org/10.1017/S1461145705005067] [PMID: 15737251]

[48] Chae BJ, Kang BJ. The effect of clozapine on blood glucose metabolism. Hum Psychopharmacol 2001; 16(3): 265-71.
[http://dx.doi.org/10.1002/hup.284] [PMID: 12404579]

[49] Alptekin K, Hafez J, Brook S, *et al.* Efficacy and tolerability of switching to ziprasidone from olanzapine, risperidone or haloperidol: An international, multicenter study. International Clinical Psychopharmacology 2009; 24(5): 229-38.

[50] von Wilmsdorff M, Bouvier M-L, Henning U, Schmitt A, Schneider-Axmann T, Gaebel W. The sex-dependent impact of chronic clozapine and haloperidol treatment on characteristics of the metabolic syndrome in a rat model. Pharmacopsychiatry 2013; 46(1): 1-9.
[PMID: 22915487]

[51] Jain S, Andridge R, Hellings JA. Loxapine for Reversal of Antipsychotic-Induced Metabolic Disturbances: A Chart Review. J Autism Dev Disord 2016; 46(4): 1344-53.
[http://dx.doi.org/10.1007/s10803-015-2675-3] [PMID: 26687568]

[52] Lin CC, Bai YM, Wang YC, *et al.* Improved body weight and metabolic outcomes in overweight or obese psychiatric patients switched to amisulpride from other atypical antipsychotics. J Clin Psychopharmacol 2009; 29(6): 529-36.
[http://dx.doi.org/10.1097/JCP.0b013e3181bf613e] [PMID: 19910716]

[53] Peuskens J, Bech P, Möller HJ, Bale R, Fleurot O, Rein W. Amisulpride *versus* risperidone in the treatment of acute exacerbations of schizophrenia. Psychiatry Res 1999; 88: 107-17.
[http://dx.doi.org/10.1016/S0165-1781(99)00075-X] [PMID: 10622347]

[54] Sechter D, Peuskens J, Fleurot O, Rein W, Lecrubier Y. Amisulpride *vs.* risperidone in chronic schizophrenia: results of a 6-month double-blind study. Neuropsychopharmacology 2002; 27(6): 1071-81.
[http://dx.doi.org/10.1016/S0893-133X(02)00375-5] [PMID: 12464464]

[55] Martin S, Ljo H, Peuskens J, *et al.* A double-blind, randomised comparative trial of amisulpride *versus* olanzapine in the treatment of schizophrenia: short-term results at two months. Curr Med Res Opin 2002; 18(6): 355-62.
[http://dx.doi.org/10.1185/030079902125001128] [PMID: 12442883]

[56] Mortimer A, Martin S, Lôo H, Peuskens J. A double-blind, randomized comparative trial of amisulpride *versus* olanzapine for 6 months in the treatment of schizophrenia. Int Clin Psychopharmacol 2004; 19(2): 63-9.
[http://dx.doi.org/10.1097/00004850-200403000-00002] [PMID: 15076013]

[57] Leucht S, Wagenpfeil S, Hamann J. Amisulpride is an atypical antipsychotic associated with low weight gain. Psychopharmacology (Berlin) 2004; 173: 112-5.
[http://dx.doi.org/10.1007/s00213-003-1721-6]

[58] Chen YY, Huang MC, Wu KC. Weight gain in a patient with schizophrenia switched from quetiapine

to amisulpride. Psychiatry Clin Neurosci 2008; 62(5): 620-1.
[http://dx.doi.org/10.1111/j.1440-1819.2008.01856.x] [PMID: 18950385]

[59]   Chang C-K, Hung GC-L. Amisulpride-associated acute onset of metabolic syndrome in a schizophrenia patient. Psychiatry Clin Neurosci 2013; 67(2): 131.
[http://dx.doi.org/10.1111/pcn.12021] [PMID: 23438173]

[60]   Olié J-P, Spina E, Murray S, Yang R. Ziprasidone and amisulpride effectively treat negative symptoms of schizophrenia: results of a 12-week, double-blind study. Int Clin Psychopharmacol 2006; 21(3): 143-51.
[http://dx.doi.org/10.1097/01.yic.0000182121.59296.70] [PMID: 16528136]

[61]   Kirchherr H, Kühn-Velten WN. Quantitative determination of forty-eight antidepressants and antipsychotics in human serum by HPLC tandem mass spectrometry: a multi-level, single-sample approach. J Chromatogr B Analyt Technol Biomed Life Sci 2006; 843(1): 100-13.
[http://dx.doi.org/10.1016/j.jchromb.2006.05.031] [PMID: 16798119]

[62]   Frahnert C, Rao ML, Grasmäder K. Analysis of eighteen antidepressants, four atypical antipsychotics and active metabolites in serum by liquid chromatography: a simple tool for therapeutic drug monitoring. J Chromatogr B Analyt Technol Biomed Life Sci 2003; 794(1): 35-47.
[http://dx.doi.org/10.1016/S1570-0232(03)00393-3] [PMID: 12888196]

[63]   Kratzsch C, Peters FT, Kraemer T, Weber AA, Maurer HH. Screening, library-assisted identification and validated quantification of fifteen neuroleptics and three of their metabolites in plasma by liquid chromatography/mass spectrometry with atmospheric pressure chemical ionization. J Mass Spectrom 2003; 38(3): 283-95.
[http://dx.doi.org/10.1002/jms.440] [PMID: 12644990]

[64]   Steuer AE, Poetzsch M, Koenig M, *et al.* Comparison of conventional liquid chromatography-tandem mass spectrometry *versus* microflow liquid chromatography-tandem mass spectrometry within the framework of full method validation for simultaneous quantification of 40 antidepressants and neuroleptics in whole blood. J Chromatogr A 2015; 1381: 87-100.
[http://dx.doi.org/10.1016/j.chroma.2014.12.084] [PMID: 25596763]

[65]   Bohbot M, Doare L, Diquet B. Determination of a new benzamide, amisulpride, in human plasma by reversed-phase ion-pair high-performance liquid chromatography. J Chromatogr A 1987; 416(2): 414-9.
[http://dx.doi.org/10.1016/0378-4347(87)80529-7] [PMID: 2886513]

[66]   Sachse J, Härtter S, Weigmann H, Hiemke C. Automated determination of amisulpride by liquid chromatography with column switching and spectrophotometric detection. J Chromatogr B Analyt Technol Biomed Life Sci 2003; 784(2): 405-10.
[http://dx.doi.org/10.1016/S1570-0232(02)00808-5] [PMID: 12505788]

[67]   Péhourcq F, Ouariki S, Bégaud B. Rapid high-performance liquid chromatographic measurement of amisulpride in human plasma: application to manage acute intoxication. J Chromatogr B Analyt Technol Biomed Life Sci 2003; 789(1): 101-5.
[http://dx.doi.org/10.1016/S1570-0232(03)00045-X] [PMID: 12726848]

[68]   Chatterjee B, Das A, Chakraborty US, Bhaumik U, Sengupta P, Pal TK. Development and validation of method for quantification of amisulpride in human plasma. Res J Biotechnol 2008; 3(SPEC. ISS.): 235-8.

[69]   Malavasi B, Locatelli M, Ripamonti M, Ascalone V. Determination of amisulpride, a new benzamide derivative, in human plasma and urine by liquid-liquid extraction or solid-phase extraction in combination with high-performance liquid chromatography and fluorescence detection. application to pharmacokinetics. J Chromatogr B Biomed Appl 1996; 676(1): 107-15.
[http://dx.doi.org/10.1016/0378-4347(95)00420-3] [PMID: 8852050]

[70]   Gschwend MH, Arnold P, Ring J, Martin W. Selective and sensitive determination of amisulpride in human plasma by liquid chromatography-tandem mass spectrometry with positive electrospray

ionisation and multiple reaction monitoring. J Chromatogr B Analyt Technol Biomed Life Sci 2006; 831(1-2): 132-9.
[http://dx.doi.org/10.1016/j.jchromb.2005.11.042] [PMID: 16386474]

[71]   Mogili R, Kanala K, Challa BR, Chandu BR, Bannoth CK. Development and validation of amisulpride in human plasma by HPLC coupled with tandem mass spectrometry and its application to a pharmacokinetic study. Sci Pharm 2011; 79(3): 583-99.
[http://dx.doi.org/10.3797/scipharm.1105-12] [PMID: 21886905]

[72]   He J, Yuan J, Du J, *et al.* Microchem J 2019; 145: 154-61.
[http://dx.doi.org/10.1016/j.microc.2018.10.029]

[73]   Papoutsis I, Rizopoulou A, Nikolaou P, Pistos C, Spiliopoulou C, Athanaselis S. A validated GC/MS method for the determination of amisulpride in whole blood. J Chromatogr B Analyt Technol Biomed Life Sci 2014; 947-948: 111-6.
[http://dx.doi.org/10.1016/j.jchromb.2013.12.024] [PMID: 24412693]

[74]   Oezkan SA, Uslu B, Sentuerk Z. Electroanalytical characteristics of amisulpride and voltammetric determination of the drug in pharmaceuticals and biological media. Electroanalysis 2004; 16: 231-7.
[http://dx.doi.org/10.1002/elan.200402828]

[75]   Ascalone V, Ripamonti M, Malavasi B. Stereospecific determination of amisulpride, a new benzamide derivative, in human plasma and urine by automated solid-phase extraction and liquid chromatography on a chiral column. application to pharmacokinetics. J Chromatogr B Biomed Appl 1996; 676(1): 95-105.
[http://dx.doi.org/10.1016/0378-4347(95)00418-1] [PMID: 8852049]

[76]   Kemp DE, Eudicone JM, McQuade RD, Chambers JS, Baker RA. Metabolic syndrome and its potential effect on treatment response to aripiprazole: a post hoc analysis of the stabilization phase of a long-term, double-blind study in patients with bipolar disorder (CN138-010). J Clin Psychopharmacol 2010; 30(5): 631-4.
[http://dx.doi.org/10.1097/JCP.0b013e3181f0569f] [PMID: 20841962]

[77]   Kemp DE, Calabrese JR, Tran QV, Pikalov A, Eudicone JM, Baker RA. Metabolic syndrome in patients enrolled in a clinical trial of aripiprazole in the maintenance treatment of bipolar I disorder: a post hoc analysis of a randomized, double-blind, placebo-controlled trial. J Clin Psychiatry 2010; 71(9): 1138-44.
[http://dx.doi.org/10.4088/JCP.09m05159gre] [PMID: 20492838]

[78]   Glick ID, Mankoski R, Eudicone JM, Marcus RN, Tran Q-V, Assunção-Talbott S. The efficacy, safety, and tolerability of aripiprazole for the treatment of schizoaffective disorder: results from a pooled analysis of a sub-population of subjects from two randomized, double-blind, placebo-controlled, pivotal trials. J Affect Disord 2009; 115(1-2): 18-26.
[http://dx.doi.org/10.1016/j.jad.2008.12.017] [PMID: 19230981]

[79]   Jakobsen KD, Bruhn CH, Pagsberg A-K, Fink-Jensen A, Nielsen J. Neurological, Metabolic, and Psychiatric Adverse Events in Children and Adolescents Treated With Aripiprazole. J Clin Psychopharmacol 2016; 36(5): 496-9.
[http://dx.doi.org/10.1097/JCP.0000000000000548] [PMID: 27504593]

[80]   Malla A, Mustafa S, Rho A, Abadi S, Lepage M, Joober R. Therapeutic effectiveness and tolerability of aripiprazole as initial choice of treatment in first episode psychosis in an early intervention service: A one-year outcome study. Schizophr Res 2016; 174(1-3): 120-5.
[http://dx.doi.org/10.1016/j.schres.2016.04.036] [PMID: 27157800]

[81]   Mankoski R, Stockton G, Manos G, *et al.* Aripiprazole treatment of irritability associated with autistic disorder and the relationship between prior antipsychotic exposure, adverse events, and weight change. J Child Adolesc Psychopharmacol 2013; 23(8): 572-6.
[http://dx.doi.org/10.1089/cap.2012.0075] [PMID: 24138011]

[82]   Findling RL, Robb A, Nyilas M, *et al.* A multiple-center, randomized, double-blind, placebo-

controlled study of oral aripiprazole for treatment of adolescents with schizophrenia. Am J Psychiatry 2008; 165(11): 1432-41.
[http://dx.doi.org/10.1176/appi.ajp.2008.07061035] [PMID: 18765484]

[83]   Muzina DJ, Momah C, Eudicone JM, *et al.* Aripiprazole monotherapy in patients with rapid-cycling bipolar I disorder: an analysis from a long-term, double-blind, placebo-controlled study. Int J Clin Pract 2008; 62(5): 679-87.
[http://dx.doi.org/10.1111/j.1742-1241.2008.01735.x] [PMID: 18373615]

[84]   Yoo HK, Joung YS, Lee J-S, *et al.* A multicenter, randomized, double-blind, placebo-controlled study of aripiprazole in children and adolescents with Tourette's disorder. J Clin Psychiatry 2013; 74(8): e772-80.
[http://dx.doi.org/10.4088/JCP.12m08189] [PMID: 24021518]

[85]   Rizzo R, Eddy CM, Calí P, Gulisano M, Cavanna AE. Metabolic effects of aripiprazole and pimozide in children with Tourette syndrome. Pediatr Neurol 2012; 47(6): 419-22.
[http://dx.doi.org/10.1016/j.pediatrneurol.2012.08.015] [PMID: 23127261]

[86]   Kishi T, Matsuda Y, Matsunaga S, Iwata N. Aripiprazole for the management of schizophrenia in the Japanese population: a systematic review and meta-analysis of randomized controlled trials. Neuropsychiatr Dis Treat 2015; 11: 419-34.
[http://dx.doi.org/10.2147/NDT.S78977] [PMID: 25759585]

[87]   Pagsberg AK, Jeppesen P, Klauber DG, *et al.* Quetiapine extended release *versus* aripiprazole in children and adolescents with first-episode psychosis: the multicentre, double-blind, randomised tolerability and efficacy of antipsychotics (TEA) trial. Lancet Psychiatry 2017; 4(8): 605-18.
[http://dx.doi.org/10.1016/S2215-0366(17)30166-9] [PMID: 28599949]

[88]   Kasteng F, Eriksson J, Sennfält K, Lindgren P. Metabolic effects and cost-effectiveness of aripiprazole *versus* olanzapine in schizophrenia and bipolar disorder. Acta Psychiatr Scand 2011; 124(3): 214-25.
[http://dx.doi.org/10.1111/j.1600-0447.2011.01716.x] [PMID: 21609324]

[89]   L'Italien GJ, Casey DE, Kan HJ, Carson WH, Marcus RN. Comparison of metabolic syndrome incidence among schizophrenia patients treated with aripiprazole *versus* olanzapine or placebo. J Clin Psychiatry 2007; 68(10): 1510-6.
[http://dx.doi.org/10.4088/JCP.v68n1006] [PMID: 17960964]

[90]   McQuade RD, Stock E, Marcus R, *et al.* A comparison of weight change during treatment with olanzapine or aripiprazole: results from a randomized, double-blind study. J Clin Psychiatry 2004; 65 (Suppl. 18): 47-56.
[PMID: 15600384]

[91]   Cohen D, Raffin M, Canitano R, *et al.* Risperidone or aripiprazole in children and adolescents with autism and/or intellectual disability: A Bayesian meta-analysis of efficacy and secondary effects. Res Autism Spectr Disord 2013; 7(1): 167-75.
[http://dx.doi.org/10.1016/j.rasd.2012.08.001]

[92]   Zhang Y, Dai G. Efficacy and metabolic influence of paliperidone ER, aripiprazole and ziprasidone to patients with first-episode schizophrenia through 52 weeks follow-up in China. Hum Psychopharmacol 2012; 27(6): 605-14.
[http://dx.doi.org/10.1002/hup.2270] [PMID: 24446539]

[93]   Lee NY, Kim SH, Jung DC, *et al.* The prevalence of metabolic syndrome in Korean patients with schizophrenia receiving a monotherapy with aripiprazole, olanzapine or risperidone. Prog Neuropsychopharmacol Biol Psychiatry 2011; 35(5): 1273-8.
[http://dx.doi.org/10.1016/j.pnpbp.2011.03.022] [PMID: 21513765]

[94]   McIntyre RS, McElroy SL, Eudicone JM, Forbes RA, Carlson BX, Baker RA. Prim Care Companion J Clin Psychiatry 2011; 13(6)

[95]   Wani RA, Dar MA, Chandel RK, *et al.* Effects of switching from olanzapine to aripiprazole on the metabolic profiles of patients with schizophrenia and metabolic syndrome: a double-blind,

randomized, open-label study. Neuropsychiatr Dis Treat 2015; 11: 685-93.
[http://dx.doi.org/10.2147/NDT.S80925] [PMID: 25792838]

[96]   Stroup TS, Byerly MJ, Nasrallah HA, *et al.* Effects of switching from olanzapine, quetiapine, and risperidone to aripiprazole on 10-year coronary heart disease risk and metabolic syndrome status: results from a randomized controlled trial. Schizophr Res 2013; 146(1-3): 190-5.
[http://dx.doi.org/10.1016/j.schres.2013.01.013] [PMID: 23434503]

[97]   De Hert M, Hanssens L, van Winkel R, *et al.* A case series: evaluation of the metabolic safety of aripiprazole. Schizophr Bull 2007; 33(3): 823-30.
[http://dx.doi.org/10.1093/schbul/sbl037] [PMID: 16940338]

[98]   Kim SH, Ivanova O, Abbasi FA, Lamendola CA, Reaven GM, Glick ID. Metabolic impact of switching antipsychotic therapy to aripiprazole after weight gain: a pilot study. J Clin Psychopharmacol 2007; 27(4): 365-8.
[http://dx.doi.org/10.1097/JCP.0b013e3180a9076c] [PMID: 17632220]

[99]   Ganguli R, Brar JS, Garbut R, Chang C-CH, Basu R. Changes in weight and other metabolic indicators in persons with schizophrenia following a switch to aripiprazole. Clin Schizophr Relat Psychoses 2011; 5(2): 75-9.
[http://dx.doi.org/10.3371/CSRP.5.2.3] [PMID: 21693430]

[100] Newcomer JW, Campos JA, Marcus RN, *et al.* A multicenter, randomized, double-blind study of the effects of aripiprazole in overweight subjects with schizophrenia or schizoaffective disorder switched from olanzapine. J Clin Psychiatry 2008; 69(7): 1046-56.
[http://dx.doi.org/10.4088/JCP.v69n0702] [PMID: 18605811]

[101] Kim S-W, Shin I-S, Kim J-M, Bae K-Y, Yang S-J, Yoon J-S. Effectiveness of switching from aripiprazole to ziprasidone in patients with schizophrenia. Clin Neuropharmacol 2010; 33(3): 121-5.
[http://dx.doi.org/10.1097/WNF.0b013e3181d52b85] [PMID: 20502130]

[102] Zheng W, Zheng Y-J, Li X-B, *et al.* Efficacy and Safety of Adjunctive Aripiprazole in Schizophrenia: Meta-Analysis of Randomized Controlled Trials. J Clin Psychopharmacol 2016; 36(6): 628-36.
[http://dx.doi.org/10.1097/JCP.0000000000000579] [PMID: 27755219]

[103] Wang L-J, Ree S-C, Huang Y-S, Hsiao C-C, Chen C-K. Adjunctive effects of aripiprazole on metabolic profiles: comparison of patients treated with olanzapine to patients treated with other atypical antipsychotic drugs. Prog Neuropsychopharmacol Biol Psychiatry 2013; 40(1): 260-6.
[http://dx.doi.org/10.1016/j.pnpbp.2012.10.010] [PMID: 23085073]

[104] Zhao J, Song X, Ai X, *et al.* Adjunctive aripiprazole treatment for risperidone-induced hyperprolactinemia: An 8-week randomized, open-label, comparative clinical trial. PLoS One 2015; 10(10)e0139717
[http://dx.doi.org/10.1371/journal.pone.0139717] [PMID: 26448615]

[105] Srisurapanont M, Suttajit S, Maneeton N, Maneeton B. Efficacy and safety of aripiprazole augmentation of clozapine in schizophrenia: a systematic review and meta-analysis of randomized-controlled trials. J Psychiatr Res 2015; 62: 38-47.
[http://dx.doi.org/10.1016/j.jpsychires.2015.01.004] [PMID: 25619176]

[106] Fleischhacker WW, Heikkinen ME, Olié J-P, *et al.* Effects of adjunctive treatment with aripiprazole on body weight and clinical efficacy in schizophrenia patients treated with clozapine: a randomized, double-blind, placebo-controlled trial. Int J Neuropsychopharmacol 2010; 13(8): 1115-25.
[http://dx.doi.org/10.1017/S1461145710000490] [PMID: 20459883]

[107] Chang JS, Lee NY, Ahn YM, Kim YS. The sustained effects of aripiprazole-augmented clozapine treatment on the psychotic symptoms and metabolic profiles of patients with refractory schizophrenia. J Clin Psychopharmacol 2012; 32(2): 282-4.
[http://dx.doi.org/10.1097/JCP.0b013e3182485871] [PMID: 22388155]

[108] Kirschbaum KM, Müller MJ, Zernig G, *et al.* Therapeutic monitoring of aripiprazole by HPLC with column-switching and spectrophotometric detection. Clin Chem 2005; 51(9): 1718-21.

[http://dx.doi.org/10.1373/clinchem.2005.049809] [PMID: 16120951]

[109] Zuo XC, Wang F, Xu P, Zhu RH, Li HD. Chromatographia 2006; 64: 387-91.
[http://dx.doi.org/10.1365/s10337-006-0037-1]

[110] Patel D, Sharma N, Patel M, Patel B, Shrivastav P, Sanyal M. Development and validation of a rapid and sensitive LC-MS/MS method for the determination of aripiprazole in human plasma: Application to a bioequivalence study. Acta Chromatogr 2014; 26(2): 203-27.
[http://dx.doi.org/10.1556/AChrom.26.2014.2.2]

[111] Ambavaram VBR, Nandigam V, Vemula M, Kalluru GR, Gajulapalle M. Liquid chromatography-tandem mass spectrometry method for simultaneous quantification of urapidil and aripiprazole in human plasma and its application to human pharmacokinetic study. Biomed Chromatogr 2013; 27(7): 916-23.
[http://dx.doi.org/10.1002/bmc.2882] [PMID: 23463771]

[112] Patel DP, Sharma P, Sanyal M, Shrivastav PS. SPE-UPLC-MS/MS method for sensitive and rapid determination of aripiprazole in human plasma to support a bioequivalence study. J Chromatogr B Analyt Technol Biomed Life Sci 2013; 925: 20-5.
[http://dx.doi.org/10.1016/j.jchromb.2013.02.022] [PMID: 23510852]

[113] Ravinder S, Bapuji AT, Mukkanti K, Raju DR, Ravikiran HLV, Reddy DC. Development and validation of an LC-ESI-MS method for quantitative determination of aripiprazole in human plasma and an application to pharmacokinetic study. J Chromatogr Sci 2012; 50(10): 893-901.
[http://dx.doi.org/10.1093/chromsci/bms087] [PMID: 22767645]

[114] Mokhtari A. Sensitive determination of aripiprazole using chemiluminescence reaction of tris(1,10-phenanthroline)ruthenium(II) with acidic Ce(IV). Anal Methods 2014; 6(24): 9588-95.
[http://dx.doi.org/10.1039/C4AY01745E]

[115] Molden E, Lunde H, Lunder N, Refsum H. Pharmacokinetic variability of aripiprazole and the active metabolite dehydroaripiprazole in psychiatric patients. Ther Drug Monit 2006; 28(6): 744-9.
[http://dx.doi.org/10.1097/01.ftd.0000249944.42859.bf] [PMID: 17164689]

[116] Kubo M, Mizooku Y, Hirao Y, Osumi T. Development and validation of an LC-MS/MS method for the quantitative determination of aripiprazole and its main metabolite, OPC-14857, in human plasma. J Chromatogr B Analyt Technol Biomed Life Sci 2005; 822(1-2): 294-9.
[http://dx.doi.org/10.1016/j.jchromb.2005.06.023] [PMID: 16005688]

[117] Huang H-C, Liu C-H, Lan T-H, *et al.* Detection and quantification of aripiprazole and its metabolite, dehydroaripiprazole, by gas chromatography-mass spectrometry in blood samples of psychiatric patients. J Chromatogr B Analyt Technol Biomed Life Sci 2007; 856(1-2): 57-61.
[http://dx.doi.org/10.1016/j.jchromb.2007.05.026] [PMID: 17602901]

[118] Wojnicz A, Belmonte C, Koller D, *et al.* Effective phospholipids removing microelution-solid phase extraction LC-MS/MS method for simultaneous plasma quantification of aripiprazole and dehydro-aripiprazole: Application to human pharmacokinetic studies. J Pharm Biomed Anal 2018; 151: 116-25.
[http://dx.doi.org/10.1016/j.jpba.2017.12.049] [PMID: 29324280]

[119] Wijma RA, van der Nagel BCH, Dierckx B, *et al.* Identification and quantification of the antipsychotics risperidone, aripiprazole, pipamperone and their major metabolites in plasma using ultra-high performance liquid chromatography-mass spectrometry. Biomed Chromatogr 2016; 30(6): 794-801.
[http://dx.doi.org/10.1002/bmc.3610] [PMID: 26447610]

[120] Dorado P, de Andrés F, Naranjo MEG, *et al.* High-performance liquid chromatography method using ultraviolet detection for the quantification of aripiprazole and dehydroaripiprazole in psychiatric patients. Drug Metabol Drug Interact 2012; 27(3): 165-70.
[http://dx.doi.org/10.1515/dmdi-2012-0016] [PMID: 23089607]

[121] Suzuki Y, Naito T, Kawakami J. Validated LC–MS/MS method for simultaneous determination of aripiprazole and its three metabolites in human plasma. Chromatographia 2017; 80(12): 1805-12.

[122] Sun Y, Lu X, Gai Y, *et al.* LC-MS/MS method for the determination of the prodrug aripiprazole lauroxil and its three metabolites in plasma and its application to *in vitro* biotransformation and animal pharmacokinetic studies. J Chromatogr B Analyt Technol Biomed Life Sci 2018; 1081-1082: 67-75.
[http://dx.doi.org/10.1016/j.jchromb.2018.02.011] [PMID: 29510329]

[123] Stepanova E, Grant B, Findling RL. Asenapine Treatment in Pediatric Patients with Bipolar I Disorder or Schizophrenia: A Review. Paediatr Drugs 2018; 20(2): 121-34.
[http://dx.doi.org/10.1007/s40272-017-0274-9] [PMID: 29170943]

[124] Bishara D, Taylor D. Asenapine monotherapy in the acute treatment of both schizophrenia and bipolar I disorder. Neuropsychiatr Dis Treat 2009; 5(1): 483-90.
[PMID: 19851515]

[125] Potkin SG, Cohen M, Panagides J. Efficacy and tolerability of asenapine in acute schizophrenia: a placebo- and risperidone-controlled trial. J Clin Psychiatry 2007; 68(10): 1492-500.
[http://dx.doi.org/10.4088/JCP.v68n1004] [PMID: 17960962]

[126] Schoemaker J, Naber D, Vrijland P, Panagides J, Emsley R. Long-term assessment of Asenapine *vs.* Olanzapine in patients with schizophrenia or schizoaffective disorder. Pharmacopsychiatry 2010; 43(4): 138-46.
[http://dx.doi.org/10.1055/s-0030-1248313] [PMID: 20205074]

[127] McIntyre RS, Cohen M, Zhao J, Alphs L, Macek TA, Panagides J. Asenapine in the treatment of acute mania in bipolar I disorder: a randomized, double-blind, placebo-controlled trial. J Affect Disord 2010; 122(1-2): 27-38.
[http://dx.doi.org/10.1016/j.jad.2009.12.028] [PMID: 20096936]

[128] McIntyre RS, Cohen M, Zhao J, Alphs L, Macek TA, Panagides J. Asenapine for long-term treatment of bipolar disorder: a double-blind 40-week extension study. J Affect Disord 2010; 126(3): 358-65.
[http://dx.doi.org/10.1016/j.jad.2010.04.005] [PMID: 20537396]

[129] Okazaki K, Yamamuro K, Kishimoto T. Reversal of olanzapine-induced weight gain in a patient with schizophrenia by switching to asenapine: a case report. Neuropsychiatr Dis Treat 2017; 13: 2837-40.
[http://dx.doi.org/10.2147/NDT.S148616] [PMID: 29200857]

[130] Orr C, Deshpande S, Sawh S, Jones PM, Vasudev K. Asenapine for the Treatment of Psychotic Disorders. Can J Psychiatry 2017; 62(2): 123-37.
[http://dx.doi.org/10.1177/0706743716661324] [PMID: 27481921]

[131] Tarazi FI, Stahl SM. Iloperidone, asenapine and lurasidone: a primer on their current status. Expert Opin Pharmacother 2012; 13(13): 1911-22.
[http://dx.doi.org/10.1517/14656566.2012.712114] [PMID: 22849428]

[132] Citrome L. Asenapine review, part II: clinical efficacy, safety and tolerability. Expert Opin Drug Saf 2014; 13(6): 803-30.
[http://dx.doi.org/10.1517/14740338.2014.908183] [PMID: 24793161]

[133] Stoner SC, Pace HA. Asenapine: a clinical review of a second-generation antipsychotic. Clin Ther 2012; 34(5): 1023-40.
[http://dx.doi.org/10.1016/j.clinthera.2012.03.002] [PMID: 22494521]

[134] Findling RL, Landbloom RL, Mackle M, *et al.* Long-term Safety of Asenapine in Pediatric Patients Diagnosed With Bipolar I Disorder: A 50-Week Open-Label, Flexible-Dose Trial. Paediatr Drugs 2016; 18(5): 367-78.
[http://dx.doi.org/10.1007/s40272-016-0184-2] [PMID: 27461426]

[135] Kane JM, Mackle M, Snow-Adami L, Zhao J, Szegedi A, Panagides J. A randomized placebo-controlled trial of asenapine for the prevention of relapse of schizophrenia after long-term treatment. J Clin Psychiatry 2011; 72(3): 349-55.
[http://dx.doi.org/10.4088/JCP.10m06306] [PMID: 21367356]

[136] de Boer T, Meulman E, Meijering H, Wieling J, Dogterom P, Lass H. Quantification of asenapine and

three metabolites in human plasma using liquid chromatography-tandem mass spectrometry with automated solid-phase extraction: application to a phase I clinical trial with asenapine in healthy male subjects. Biomed Chromatogr 2012; 26(2): 156-65.
[http://dx.doi.org/10.1002/bmc.1640] [PMID: 21557265]

[137] de Boer T, Meulman E, Meijering H, Wieling J, Dogterom P, Lass H. Development and validation of automated SPE-HPLC-MS/MS methods for the quantification of asenapine, a new antipsychotic agent, and its two major metabolites in human urine. Biomed Chromatogr 2012; 26(12): 1461-3.
[http://dx.doi.org/10.1002/bmc.2722] [PMID: 22344545]

[138] Sistik P, Urinovska R, Brozmanova H, Kacirova I, Silhan P, Lemr K. Fast simultaneous LC/MS/MS determination of 10 active compounds in human serum for therapeutic drug monitoring in psychiatric medication. Biomed Chromatogr 2016; 30(2): 217-24.
[http://dx.doi.org/10.1002/bmc.3538] [PMID: 26094602]

[139] Montenarh D, Hopf M, Maurer HH, Schmidt P, Ewald AH. Development and validation of a multi-analyte LC-MS/MS approach for quantification of neuroleptics in whole blood, plasma, and serum. Drug Test Anal 2016; 8(10): 1080-9.
[http://dx.doi.org/10.1002/dta.1923] [PMID: 26607679]

[140] Patel NP, Sanyal M, Sharma N, Patel DS, Shrivastav PS, Patel BN. Determination of asenapine in presence of its inactive metabolites in human plasma by LC-MS/MS. J Pharm Anal 2018; 8(5): 341-7.
[http://dx.doi.org/10.1016/j.jpha.2018.06.002] [PMID: 30345149]

[141] Miller C, Pleitez O, Anderson D, Mertens-Maxham D, Wade N. Asenapine (Saphris®): GC-MS method validation and the postmortem distribution of a new atypical antipsychotic medication. J Anal Toxicol 2013; 37(8): 559-64.
[http://dx.doi.org/10.1093/jat/bkt076] [PMID: 24009049]

[142] Kovatsi L, Manousi N, Papadoyannis I. Direct UHPLC-DAD Method to Determine Asenapine, Paroxetine and Fluvoxamine in Human Blood Serum, Urine and Cerebrospinal Fluid. Curr Pharm Anal 2016; 12(4): 349-56.
[http://dx.doi.org/10.2174/1573412912666151217182814]

[143] Kovatsi L, Titopoulou A, Tsakalof A, Samanidou V. HPLC Analysis of Antipsychotic Asenapine in Alternative Biomatrices: Hair and Nail Clippings. J Liq Chromatogr Relat Technol 2015; 38(18): 1666-70.
[http://dx.doi.org/10.1080/10826076.2015.1089894]

[144] Protti M, Vignali A, Sanchez Blanco T, *et al.* Enantioseparation and determination of asenapine in biological fluid micromatrices by HPLC with diode array detection. J Sep Sci 2018; 41(6): 1257-65.
[http://dx.doi.org/10.1002/jssc.201701315] [PMID: 29266728]

[145] Protti M, Mandrioli R, Mercolini L. Tutorial: Volumetric absorptive microsampling (VAMS). Anal Chim Acta 2019; 1046: 32-47.
[http://dx.doi.org/10.1016/j.aca.2018.09.004] [PMID: 30482302]

[146] Siskind D, Friend N, Russell A, *et al.* CoMET: a protocol for a randomised controlled trial of co-commencement of METformin as an adjunctive treatment to attenuate weight gain and metabolic syndrome in patients with schizophrenia newly commenced on clozapine. BMJ Open 2018; 8(3)e021000
[http://dx.doi.org/10.1136/bmjopen-2017-021000] [PMID: 29500217]

[147] Lally J, Gallagher A, Bainbridge E, Avalos G, Ahmed M, McDonald C. Increases in triglyceride levels are associated with clinical response to clozapine treatment. J Psychopharmacol (Oxford) 2013; 27(4): 401-3.
[http://dx.doi.org/10.1177/0269881112472568] [PMID: 23325369]

[148] Bai YM, Lin C-C, Chen J-Y, Chen TT, Su T-P, Chou P. Association of weight gain and metabolic syndrome in patients taking clozapine: an 8-year cohort study. J Clin Psychiatry 2011; 72(6): 751-6.
[http://dx.doi.org/10.4088/JCP.09m05402yel] [PMID: 21208584]

[149] Malhi G, Adams D, Plain J, Coulston C, Herman M, Walter G. Clozapine and cardiometabolic health in chronic schizophrenia: correlations and consequences in a clinical context. Australas Psychiatry 2010; 18(1): 32-41.
[http://dx.doi.org/10.3109/10398560903254193] [PMID: 20039791]

[150] Grover S, Hazari N, Chakrabarti S, Avasthi A. Metabolic Disturbances, Side Effect Profile and Effectiveness of Clozapine in Adolescents. Indian J Psychol Med 2016; 38(3): 224-33.
[http://dx.doi.org/10.4103/0253-7176.183091] [PMID: 27335518]

[151] Lamberti JS, Olson D, Crilly JF, *et al.* Prevalence of the metabolic syndrome among patients receiving clozapine. Am J Psychiatry 2006; 163(7): 1273-6.
[http://dx.doi.org/10.1176/ajp.2006.163.7.1273] [PMID: 16816234]

[152] Brunero S, Lamont S, Fairbrother G. Prevalence and predictors of metabolic syndrome among patients attending an outpatient clozapine clinic in Australia. Arch Psychiatr Nurs 2009; 23(3): 261-8.
[http://dx.doi.org/10.1016/j.apnu.2008.06.007] [PMID: 19446781]

[153] Josiassen RC, Filmyer DM, Curtis JL, *et al.* An archival, follow-forward exploration of the metabolic syndrome in randomly selected, clozapine-treated patients. Clin Schizophr Relat Psychoses 2009; 3(2): 87-96.
[http://dx.doi.org/10.3371/CSRP.3.2.3]

[154] Howes OD, Bhatnagar A, Gaughran FP, Amiel SA, Murray RM, Pilowsky LS. A prospective study of impairment in glucose control caused by clozapine without changes in insulin resistance. Am J Psychiatry 2004; 161(2): 361-3.
[http://dx.doi.org/10.1176/appi.ajp.161.2.361] [PMID: 14754788]

[155] Suetani RJ, Siskind D, Reichhold H, Kisely S. Genetic variants impacting metabolic outcomes among people on clozapine: a systematic review and meta-analysis. Psychopharmacology (Berl) 2017; 234(20): 2989-3008.
[http://dx.doi.org/10.1007/s00213-017-4728-0] [PMID: 28879574]

[156] Lu M-L, Chen T-T, Kuo P-H, Hsu C-C, Chen C-H. Effects of adjunctive fluvoxamine on metabolic parameters and psychopathology in clozapine-treated patients with schizophrenia: A 12-week, randomized, double-blind, placebo-controlled study. Schizophr Res 2018; 193: 126-33.
[http://dx.doi.org/10.1016/j.schres.2017.06.030] [PMID: 28688742]

[157] Polcwiartek C, Nielsen J. The clinical potentials of adjunctive fluvoxamine to clozapine treatment: a systematic review. Psychopharmacology (Berl) 2016; 233(5): 741-50.
[http://dx.doi.org/10.1007/s00213-015-4161-1] [PMID: 26626327]

[158] Zimbron J, Khandaker GM, Toschi C, Jones PB, Fernandez-Egea E. A systematic review and meta-analysis of randomised controlled trials of treatments for clozapine-induced obesity and metabolic syndrome. Eur Neuropsychopharmacol 2016; 26(9): 1353-65.
[http://dx.doi.org/10.1016/j.euroneuro.2016.07.010] [PMID: 27496573]

[159] Siskind DJ, Leung J, Russell AW, Wysoczanski D, Kisely S. Metformin for clozapine associated obesity: A systematic review and meta-analysis. PLoS One 2016; 11(6)e0156208
[http://dx.doi.org/10.1371/journal.pone.0156208] [PMID: 27304831]

[160] Carrizo E, Fernández V, Connell L, *et al.* Extended release metformin for metabolic control assistance during prolonged clozapine administration: a 14 week, double-blind, parallel group, placebo-controlled study. Schizophr Res 2009; 113(1): 19-26.
[http://dx.doi.org/10.1016/j.schres.2009.05.007] [PMID: 19515536]

[161] Chen C-H, Huang M-C, Kao C-F, *et al.* Effects of adjunctive metformin on metabolic traits in nondiabetic clozapine-treated patients with schizophrenia and the effect of metformin discontinuation on body weight: a 24-week, randomized, double-blind, placebo-controlled study. J Clin Psychiatry 2013; 74(5): e424-30.
[http://dx.doi.org/10.4088/JCP.12m08186] [PMID: 23759461]

[162] Hebrani P, Manteghi AA, Behdani F, *et al.* Double-blind, randomized, clinical trial of metformin as add-on treatment with clozapine in treatment of schizophrenia disorder. J Res Med Sci 2015; 20(4): 364-71.
[PMID: 26109992]

[163] Henderson DC, Fan X, Sharma B, *et al.* A double-blind, placebo-controlled trial of rosiglitazone for clozapine-induced glucose metabolism impairment in patients with schizophrenia. Acta Psychiatr Scand 2009; 119(6): 457-65.
[http://dx.doi.org/10.1111/j.1600-0447.2008.01325.x] [PMID: 19183127]

[164] Sacchetti E, Galluzzo A, Valsecchi P, Romeo F, Gorini B, Warrington L. Ziprasidone *vs.* clozapine in schizophrenia patients refractory to multiple antipsychotic treatments: the MOZART study. Schizophr Res 2009; 113(1): 112-21.
[http://dx.doi.org/10.1016/j.schres.2009.05.002] [PMID: 19606529]

[165] Chukwuma J, Morgan D, Sargeant M, Hughes G. A cross-sectional survey of the prevalence of metabolic syndrome in adults with serious mental illness treated with olanzapine or clozapine. Prim Care Community Psychiatry 2008; 13(2): 53-8.

[166] Meltzer HY, Bobo WV, Roy A, *et al.* A randomized, double-blind comparison of clozapine and high-dose olanzapine in treatment-resistant patients with schizophrenia. J Clin Psychiatry 2008; 69(2): 274-85.
[http://dx.doi.org/10.4088/JCP.v69n0214] [PMID: 18232726]

[167] De Vivo S, Zuccaro A, Ventimiglia A. Benefit of aripiprazole in association with a low dose of clozapine in a case of metabolic syndrome. Italian Journal of Psychopathology 2011; 17(1): 158-60.

[168] Nielsen J, Emborg C, Gydesen S, *et al.* Augmenting clozapine with sertindole: a double-blind, randomized, placebo-controlled study. J Clin Psychopharmacol 2012; 32(2): 173-8.
[http://dx.doi.org/10.1097/JCP.0b013e318248dfb8] [PMID: 22367659]

[169] Honer WG, Thornton AE, Chen EYH, *et al.* Clozapine alone *versus* clozapine and risperidone with refractory schizophrenia. N Engl J Med 2006; 354(5): 472-82.
[http://dx.doi.org/10.1056/NEJMoa053222] [PMID: 16452559]

[170] Kemp DE, De Hert M, Rahman Z, *et al.* Investigation into the long-term metabolic effects of aripiprazole adjunctive to lithium, valproate, or lamotrigine. J Affect Disord 2013; 148(1): 84-91.
[http://dx.doi.org/10.1016/j.jad.2012.11.054] [PMID: 23261129]

[171] Richter K. Determination of clozapine in human serum by capillary gas chromatography. J Chromatogr A 1988; 434(2): 465-8.
[http://dx.doi.org/10.1016/S0378-4347(88)80014-8] [PMID: 3246536]

[172] Heipertz R, Pilz H, Beckers W. Serum concentrations of clozapine determined by nitrogen selective gas chromatography. Arch Toxicol 1977; 37(4): 313-8.
[http://dx.doi.org/10.1007/BF00330823] [PMID: 578707]

[173] Bondesson U, Lindström LH. Determination of clozapine and its N-demethylated metabolite in plasma by use of gas chromatography-mass spectrometry with single ion detection. Psychopharmacology (Berl) 1988; 95(4): 472-5.
[http://dx.doi.org/10.1007/BF00172957] [PMID: 3145517]

[174] Olesen OV, Poulsen B. On-line fully automated determination of clozapine and desmethylclozapine in human serum by solid-phase extraction on exchangeable cartridges and liquid chromatography using a methanol buffer mobile phase on unmodified silica. J Chromatogr A 1993; 622: 39-46.
[http://dx.doi.org/10.1016/0378-4347(93)80247-2]

[175] Wang ZR, Lu ML, Xu PP, Zeng YL, Zeng YL. Determination of clozapine and its metabolites in serum and urine by reversed phase HPLC. Biomed Chromatogr 1986; 1(2): 53-7.
[http://dx.doi.org/10.1002/bmc.1130010203] [PMID: 3507212]

[176] Haring C, Humpel C, Auer B, *et al.* Clozapine plasma levels determined by high-performance liquid

chromatography with ultraviolet detection. J Chromatogr A 1988; 428(1): 160-6.
[http://dx.doi.org/10.1016/S0378-4347(00)83902-X] [PMID: 3170669]

[177]   Chovan JP, Vermeulen JD. High-performance liquid chromatographic method for a clozapine analogue, CGS 13429, and its N-oxide and desmethyl metabolites. J Chromatogr A 1989; 494: 413-9.
[http://dx.doi.org/10.1016/S0378-4347(00)82697-3]

[178]   Wilhelm D, Kemper A. High-performance liquid chromatographic procedure for the determination of clozapine, haloperidol, droperidol and several benzodiazepines in plasma. J Chromatogr A 1990; 525: 218-24.
[http://dx.doi.org/10.1016/S0378-4347(00)83396-4]

[179]   Chung MC, Lin SK, Chang WH, Jann MW. Determination of clozapine and desmethylclozapine in human plasma by high-performance liquid chromatography with ultraviolet detection. J Chromatogr A 1993; 613(1): 168-73.
[http://dx.doi.org/10.1016/0378-4347(93)80212-M] [PMID: 8458896]

[180]   Volpicelli SA, Centorrino F, Puopolo PR, *et al.* Determination of clozapine, norclozapine, and clozapine-N-oxide in serum by liquid chromatography. Clin Chem 1993; 39(8): 1656-9.
[PMID: 8353952]

[181]   Weigmann H, Hiemke C. Determination of clozapine and its major metabolites in human serum using automated solid-phase extraction and subsequent isocratic high-performance liquid chromatography with ultraviolet detection. J Chromatogr A 1992; 583(2): 209-16.
[http://dx.doi.org/10.1016/0378-4347(92)80554-4] [PMID: 1478985]

[182]   Fadiran EO, Leslie J, Fossler M, Young D. Determination of clozapine and its major metabolites in human serum and rat plasma by liquid chromatography using solid-phase extraction and ultraviolet detection. J Pharm Biomed Anal 1995; 13(2): 185-90.
[http://dx.doi.org/10.1016/0731-7085(93)E0020-N] [PMID: 7766727]

[183]   Gupta RN. Column liquid chromatographic determination of clozapine and N-desmethylclozapine in human serum using solid-phase extraction. J Chromatogr B Biomed Appl 1995; 673(2): 311-5.
[http://dx.doi.org/10.1016/0378-4347(95)00262-3] [PMID: 8611967]

[184]   Hariharan U, Hariharan M, Naickar JS, Tandon R. Determination of Clozapine and Its Two Major Metabolites in Human Serum by Liquid Chromatography Using Ultraviolet Detection. J Liq Chromatogr Relat Technol 1996; 19: 2409-17.
[http://dx.doi.org/10.1080/10826079608014026]

[185]   Edno L, Combourieu I, Cazenave M, Tignol J. Assay for quantitation of clozapine and its metabolite N-desmethylclozapine in human plasma by high-performance liquid chromatography with ultraviolet detection. J Pharm Biomed Anal 1997; 16(2): 311-8.
[http://dx.doi.org/10.1016/S0731-7085(97)00048-4] [PMID: 9408849]

[186]   Liu YY, van Troostwijk LJ, Guchelaar HJ. Simultaneous determination of clozapine, norclozapine and clozapine-N-oxide in human plasma by high-performance liquid chromatography with ultraviolet detection. Biomed Chromatogr 2001; 15(4): 280-6.
[http://dx.doi.org/10.1002/bmc.73] [PMID: 11438972]

[187]   Weigmann H, Härtter S, Maehrlein S, *et al.* Simultaneous determination of olanzapine, clozapine and demethylated metabolites in serum by on-line column-switching high-performance liquid chromatography. J Chromatogr B Biomed Sci Appl 2001; 759(1): 63-71.
[http://dx.doi.org/10.1016/S0378-4347(01)00215-8] [PMID: 11499630]

[188]   Palego L, Dell'Osso L, Marazziti D, *et al.* Simultaneous analysis of clozapine, clomipramine and their metabolites by reversed-phase liquid chromatography. Prog Neuropsychopharmacol Biol Psychiatry 2001; 25(3): 519-33.
[http://dx.doi.org/10.1016/S0278-5846(00)00184-6] [PMID: 11370995]

[189]   Madej K, Biedroń A, Garbacik A. Study of separation and extraction conditions for five neuroleptic drugs by an LLE-HPLC-DAD method in human plasma. J Liq Chromatogr Relat Technol 2009;

32(20): 3025-37.
[http://dx.doi.org/10.1080/10826070903320582]

[190] Humpel C, Haring C, Saria A. Rapid and sensitive determination of clozapine in human plasma using high-performance liquid chromatography and amperometric detection. J Chromatogr A 1989; 491(1): 235-9.
[http://dx.doi.org/10.1016/S0378-4347(00)82838-8] [PMID: 2793975]

[191] Schulz E, Fleischhaker C, Remschmidt H. Determination of clozapine and its major metabolites in serum samples of adolescent schizophrenic patients by high-performance liquid chromatography. Data from a prospective clinical trial. Pharmacopsychiatry 1995; 28(1): 20-5.
[http://dx.doi.org/10.1055/s-2007-979583] [PMID: 7746841]

[192] Raggi MA, Bugamelli F, Mandrioli R, De Ronchi D, Volterra V. Determination of clozapine in human plasma by solid-phase extraction and liquid chromatography with amperometric detection. Chromatographia 1998; 47: 8-12.
[http://dx.doi.org/10.1007/BF02466779]

[193] Raggi MA, Bugamelli F, Mandrioli R, De Ronchi D, Volterra V. Development and validation of an HPLC method for the simultaneous determination of clozapine and desmethylclozapine in plasma of schizophrenic patients. Chromatographia 1999; 49: 75-80.
[http://dx.doi.org/10.1007/BF02467191]

[194] Raggi MA, Bugamelli F, Mandrioli R, Sabbioni C, Volterra V, Fanali S. Rapid capillary electrophoretic method for the determination of clozapine and desmethylclozapine in human plasma. J Chromatogr A 2001; 916(1-2): 289-96.
[http://dx.doi.org/10.1016/S0021-9673(01)00520-9] [PMID: 11382303]

[195] Aravagiri M, Marder SR. Simultaneous determination of clozapine and its N-desmethyl and N-oxide metabolites in plasma by liquid chromatography/electrospray tandem mass spectrometry and its application to plasma level monitoring in schizophrenic patients. J Pharm Biomed Anal 2001; 26(2): 301-11.
[http://dx.doi.org/10.1016/S0731-7085(01)00410-1] [PMID: 11470207]

[196] Kollroser M, Schober C. Direct-injection high performance liquid chromatography ion trap mass spectrometry for the quantitative determination of olanzapine, clozapine and N-desmethylclozapine in human plasma. Rapid Commun Mass Spectrom 2002; 16(13): 1266-72.
[http://dx.doi.org/10.1002/rcm.718] [PMID: 12112253]

[197] Uřinovská R, Brozmanová H, Šištík P, et al. Liquid chromatography-tandem mass spectrometry method for determination of five antidepressants and four atypical antipsychotics and their main metabolites in human serum. J Chromatogr B Analyt Technol Biomed Life Sci 2012; 907: 101-7.
[http://dx.doi.org/10.1016/j.jchromb.2012.09.009] [PMID: 23026228]

[198] Couchman L, Subramaniam K, Fisher DS, Belsey SL, Handley SA, Flanagan RJ. Automated Analysis of Clozapine and Norclozapine in Human Plasma Using Novel Extraction Plate Technology and Flow-Injection Tandem Mass Spectrometry. Ther Drug Monit 2016; 38(1): 42-9.
[http://dx.doi.org/10.1097/FTD.0000000000000231] [PMID: 26349082]

[199] Couchman L, Belsey SL, Handley SA, Flanagan RJ. A novel approach to quantitative LC-MS/MS: therapeutic drug monitoring of clozapine and norclozapine using isotopic internal calibration. Anal Bioanal Chem 2013; 405(29): 9455-66.
[http://dx.doi.org/10.1007/s00216-013-7361-8] [PMID: 24091736]

[200] Citrome L, Weiden PJ, Alva G, et al. Switching to iloperidone: An omnibus of clinically relevant observations from a 12-week, open-label, randomized clinical trial in 500 persons with schizophrenia. Clin Schizophr Relat Psychoses 2015; 8(4): 183-95.
[http://dx.doi.org/10.3371/CSRP.CIWE.103114] [PMID: 25367165]

[201] Cutler AJ, Kalali AH, Mattingly GW, Kunovac J, Meng X. Long-term safety and tolerability of iloperidone: results from a 25-week, open-label extension trial. CNS Spectr 2013; 18(1): 43-54.

[http://dx.doi.org/10.1017/S1092852912000764] [PMID: 23312567]

[202] Cutler AJ, Kalali AH, Weiden PJ, Hamilton J, Wolfgang CD. Four-week, double-blind, placebo- and ziprasidone-controlled trial of iloperidone in patients with acute exacerbations of schizophrenia. J Clin Psychopharmacol 2008; 28(2) (Suppl. 1): S20-8.
[http://dx.doi.org/10.1097/JCP.0b013e318169d4ce] [PMID: 18334909]

[203] Kane JM, Lauriello J, Laska E, Di Marino M, Wolfgang CD. Long-term efficacy and safety of iloperidone: results from 3 clinical trials for the treatment of schizophrenia. J Clin Psychopharmacol 2008; 28(2) (Suppl. 1): S29-35.
[http://dx.doi.org/10.1097/JCP.0b013e318169cca7] [PMID: 18334910]

[204] McEvoy JP, Lieberman JA, Perkins DO, *et al.* Efficacy and tolerability of olanzapine, quetiapine, and risperidone in the treatment of early psychosis: a randomized, double-blind 52-week comparison. Am J Psychiatry 2007; 164(7): 1050-60.
[http://dx.doi.org/10.1176/ajp.2007.164.7.1050] [PMID: 17606657]

[205] Weiden PJ, Cutler AJ, Polymeropoulos MH, Wolfgang CD. Safety profile of iloperidone: a pooled analysis of 6-week acute-phase pivotal trials. J Clin Psychopharmacol 2008; 28(2) (Suppl. 1): S12-9.
[http://dx.doi.org/10.1097/JCP.0b013e3181694f5a] [PMID: 18334908]

[206] Mutlib AE, Strupczewski JT. Picogram determination of iloperidone in human plasma by solid-phase extraction and by high-performance liquid chromatography-selected-ion monitoring electrospray mass spectrometry. J Chromatogr B Biomed Appl 1995; 669(2): 237-46.
[http://dx.doi.org/10.1016/0378-4347(95)00114-X] [PMID: 7581900]

[207] Jia M, Li J, He X, *et al.* Simultaneous determination of iloperidone and its two active metabolites in human plasma by liquid chromatography-tandem mass spectrometry: application to a pharmacokinetic study. J Chromatogr B Analyt Technol Biomed Life Sci 2013; 928: 52-7.
[http://dx.doi.org/10.1016/j.jchromb.2013.03.026] [PMID: 23618742]

[208] Parekh JM, Sanyal M, Yadav M, Shrivastav PS. Stable-isotope dilution LC-MS/MS assay for determination of iloperidone and its two major metabolites, P 88 and P 95, in human plasma: application to a bioequivalence study. Bioanalysis 2013; 5(6): 669-86.
[http://dx.doi.org/10.4155/bio.13.25] [PMID: 23484785]

[209] Mao J-J, Zhang Q-Y, Hua W-Y. Simultaneous determination of iloperidone and its two active metabolites in human plasma by LC-MS/MS. Zhongguo Xin Yao Zazhi 2015; 24(4): 430-8.

[210] Mutlib AE, Strupczewski JT, Chesson SM. Application of hyphenated LC/NMR and LC/MS techniques in rapid identification of *in vitro* and *in vivo* metabolites of iloperidone. Drug Metab Dispos 1995; 23(9): 951-64.
[PMID: 8565786]

[211] Mutlib AE, Klein JT. Application of liquid chromatography/mass spectrometry in accelerating the identification of human liver cytochrome P450 isoforms involved in the metabolism of iloperidone. J Pharmacol Exp Ther 1998; 286(3): 1285-93.
[PMID: 9732390]

[212] Ansermot N, Brawand-Amey M, Kottelat A, Eap CB. Fast quantification of ten psychotropic drugs and metabolites in human plasma by ultra-high performance liquid chromatography tandem mass spectrometry for therapeutic drug monitoring. J Chromatogr A 2013; 1292: 160-72.
[http://dx.doi.org/10.1016/j.chroma.2012.12.071] [PMID: 23380367]

[213] Patteet L, Maudens KE, Sabbe B, Morrens M, De Doncker M, Neels H. High throughput identification and quantification of 16 antipsychotics and 8 major metabolites in serum using ultra-high performance liquid chromatography-tandem mass spectrometry. Clin Chim Acta 2014; 429: 51-8.
[http://dx.doi.org/10.1016/j.cca.2013.11.024] [PMID: 24291056]

[214] Patteet L, Maudens KE, Stove CP, *et al.* The use of dried blood spots for quantification of 15 antipsychotics and 7 metabolites with ultra-high performance liquid chromatography - tandem mass spectrometry. Drug Testing & Analysis 2014; 7: 502-11.

[215] Zheng W, Cai D-B, Yang X-H, *et al.* Short-term efficacy and tolerability of lurasidone in the treatment of acute schizophrenia: A meta-analysis of randomized controlled trials. J Psychiatr Res 2018; 103: 244-51.
[http://dx.doi.org/10.1016/j.jpsychires.2018.06.005] [PMID: 29906709]

[216] Calabrese JR, Pikalov A, Streicher C, Cucchiaro J, Mao Y, Loebel A. Lurasidone in combination with lithium or valproate for the maintenance treatment of bipolar I disorder. Eur Neuropsychopharmacol 2017; 27(9): 865-76.
[http://dx.doi.org/10.1016/j.euroneuro.2017.06.013] [PMID: 28689688]

[217] Pikalov A, Tsai J, Mao Y, Silva R, Cucchiaro J, Loebel A. Long-term use of lurasidone in patients with bipolar disorder: safety and effectiveness over 2 years of treatment. Int J Bipolar Disord 2017; 5(1): 9.
[http://dx.doi.org/10.1186/s40345-017-0075-7] [PMID: 28168632]

[218] Goldman R, Loebel A, Cucchiaro J, Deng L, Findling RL. Efficacy and Safety of Lurasidone in Adolescents with Schizophrenia: A 6-Week, Randomized Placebo-Controlled Study. J Child Adolesc Psychopharmacol 2017; 27(6): 516-25.
[http://dx.doi.org/10.1089/cap.2016.0189] [PMID: 28475373]

[219] Ogasa M, Kimura T, Nakamura M, Guarino J. Lurasidone in the treatment of schizophrenia: a 6-week, placebo-controlled study. Psychopharmacology (Berl) 2013; 225(3): 519-30.
[http://dx.doi.org/10.1007/s00213-012-2838-2] [PMID: 22903391]

[220] Nakamura M, Ogasa M, Guarino J, *et al.* Lurasidone in the treatment of acute schizophrenia: a double-blind, placebo-controlled trial. J Clin Psychiatry 2009; 70(6): 829-36.
[http://dx.doi.org/10.4088/JCP.08m04905] [PMID: 19497249]

[221] Loebel A, Cucchiaro J, Xu J, Sarma K, Pikalov A, Kane JM. Effectiveness of lurasidone *vs.* quetiapine XR for relapse prevention in schizophrenia: a 12-month, double-blind, noninferiority study. Schizophr Res 2013; 147(1): 95-102.
[http://dx.doi.org/10.1016/j.schres.2013.03.013] [PMID: 23583011]

[222] Potkin SG, Ogasa M, Cucchiaro J, Loebel A. Double-blind comparison of the safety and efficacy of lurasidone and ziprasidone in clinically stable outpatients with schizophrenia or schizoaffective disorder. Schizophr Res 2011; 132(2-3): 101-7.
[http://dx.doi.org/10.1016/j.schres.2011.04.008] [PMID: 21889878]

[223] Citrome L, Weiden PJ, McEvoy JP, *et al.* Effectiveness of lurasidone in schizophrenia or schizoaffective patients switched from other antipsychotics: a 6-month, open-label, extension study. CNS Spectr 2014; 19(4): 330-9.
[http://dx.doi.org/10.1017/S109285291300093X] [PMID: 24330868]

[224] Katteboina MY, Pilli NR, Mullangi R, Seelam RR, Satla SR. LC-MS/MS assay for the determination of lurasidone and its active metabolite, ID-14283 in human plasma and its application to a clinical pharmacokinetic study. Biomed Chromatogr 2016; 30(7): 1065-74.
[http://dx.doi.org/10.1002/bmc.3651] [PMID: 26577488]

[225] Chae YJ, Koo TS, Lee KR. A sensitive and selective LC-MS method for the determination of lurasidone in rat plasma, bile, and urine. Chromatographia 2012; 75: 1117-28.
[http://dx.doi.org/10.1007/s10337-012-2294-5]

[226] Koo TS, Kim SJ, Lee J, Ha DJ, Baek M, Moon H. Quantification of lurasidone, an atypical antipsychotic drug, in rat plasma with high-performance liquid chromatography with tandem mass spectrometry. Biomed Chromatogr 2011; 25(12): 1389-94.
[http://dx.doi.org/10.1002/bmc.1625] [PMID: 21387355]

[227] Salviato Balbão M, Cecílio Hallak JE, Arcoverde Nunes E, *et al.* Olanzapine, weight change and metabolic effects: a naturalistic 12-month follow up. Ther Adv Psychopharmacol 2014; 4(1): 30-6.
[http://dx.doi.org/10.1177/2045125313507738] [PMID: 24490028]

[228] Citrome L, Holt RIG, Walker DJ, Hoffmann VP. Weight gain and changes in metabolic variables following olanzapine treatment in schizophrenia and bipolar disorder. Clin Drug Investig 2011; 31(7): 455-82.
[http://dx.doi.org/10.2165/11589060-000000000-00000] [PMID: 21495734]

[229] Chiu C-C, Chen C-H, Chen B-Y, Yu S-H, Lu M-L. The time-dependent change of insulin secretion in schizophrenic patients treated with olanzapine. Prog Neuropsychopharmacol Biol Psychiatry 2010; 34(6): 866-70.
[http://dx.doi.org/10.1016/j.pnpbp.2010.04.003] [PMID: 20394794]

[230] Flank J, Sung L, Dvorak CC, Spettigue W, Dupuis LL. The safety of olanzapine in young children: a systematic review and meta-analysis. Drug Saf 2014; 37(10): 791-804.
[http://dx.doi.org/10.1007/s40264-014-0219-y] [PMID: 25145624]

[231] Meyer JM, Rosenblatt LC, Kim E, Baker RA, Whitehead R. The moderating impact of ethnicity on metabolic outcomes during treatment with olanzapine and aripiprazole in patients with schizophrenia. J Clin Psychiatry 2009; 70(3): 318-25.
[http://dx.doi.org/10.4088/JCP.08m04267] [PMID: 19192469]

[232] Stauffer VL, Sniadecki JL, Piezer KW, *et al.* Impact of race on efficacy and safety during treatment with olanzapine in schizophrenia, schizophreniform or schizoaffective disorder. BMC Psychiatry 2010; 10: 89.
[http://dx.doi.org/10.1186/1471-244X-10-89] [PMID: 21047395]

[233] Dabbert D, Heinze M. Case report: Reversible metabolic syndrome under olanzapine. Psychopharmakotherapie 2009; 16(1): 32-3.

[234] Silverman BL, Martin W, Memisoglu A, DiPetrillo L, Correll CU, Kane JM. A randomized, double-blind, placebo-controlled proof of concept study to evaluate samidorphan in the prevention of olanzapine-induced weight gain in healthy volunteers. Schizophr Res 2018; 195: 245-51.
[http://dx.doi.org/10.1016/j.schres.2017.10.014] [PMID: 29158012]

[235] Taveira TH, Wu W-C, Tschibelu E, *et al.* The effect of naltrexone on body fat mass in olanzapine-treated schizophrenic or schizoaffective patients: a randomized double-blind placebo-controlled pilot study. J Psychopharmacol (Oxford) 2014; 28(4): 395-400.
[http://dx.doi.org/10.1177/0269881113509904] [PMID: 24218048]

[236] Holka-Pokorska JA, Radzio R, Jarema M, Wichniak A. [The stabilizing effect of dehydroepiandrosterone on clinical parameters of metabolic syndrome in patients with schizophrenia treated with olanzapine - a randomized, double-blind trial]. Psychiatr Pol 2015; 49(2): 363-76.
[http://dx.doi.org/10.12740/PP/30180] [PMID: 26093599]

[237] Praharaj SK, Jana AK, Goyal N, Sinha VK. Metformin for olanzapine-induced weight gain: a systematic review and meta-analysis. Br J Clin Pharmacol 2011; 71(3): 377-82.
[http://dx.doi.org/10.1111/j.1365-2125.2010.03783.x] [PMID: 21284696]

[238] Baptista T, Uzcátegui E, Rangel N, *et al.* Metformin plus sibutramine for olanzapine-associated weight gain and metabolic dysfunction in schizophrenia: a 12-week double-blind, placebo-controlled pilot study. Psychiatry Res 2008; 159(1-2): 250-3.
[http://dx.doi.org/10.1016/j.psychres.2008.01.011] [PMID: 18374423]

[239] Chen C-H, Chiu C-C, Huang M-C, Wu T-H, Liu H-C, Lu M-L. Metformin for metabolic dysregulation in schizophrenic patients treated with olanzapine. Prog Neuropsychopharmacol Biol Psychiatry 2008; 32(4): 925-31.
[http://dx.doi.org/10.1016/j.pnpbp.2007.11.013] [PMID: 18082302]

[240] Narula PK, Rehan HS, Unni KES, Gupta N. Topiramate for prevention of olanzapine associated weight gain and metabolic dysfunction in schizophrenia: a double-blind, placebo-controlled trial. Schizophr Res 2010; 118(1-3): 218-23.
[http://dx.doi.org/10.1016/j.schres.2010.02.001] [PMID: 20207521]

[241]   Fadai F, Mousavi B, Ashtari Z, *et al.* Saffron aqueous extract prevents metabolic syndrome in patients with schizophrenia on olanzapine treatment: a randomized triple blind placebo controlled study. Pharmacopsychiatry 2014; 47(4-5): 156-61.
[http://dx.doi.org/10.1055/s-0034-1382001] [PMID: 24955550]

[242]   Modabbernia A, Heidari P, Soleimani R, *et al.* Melatonin for prevention of metabolic side-effects of olanzapine in patients with first-episode schizophrenia: randomized double-blind placebo-controlled study. J Psychiatr Res 2014; 53(1): 133-40.
[http://dx.doi.org/10.1016/j.jpsychires.2014.02.013] [PMID: 24607293]

[243]   Mostafavi A, Solhi M, Mohammadi M-R, Hamedi M, Keshavarzi M, Akhondzadeh S. Melatonin decreases olanzapine induced metabolic side-effects in adolescents with bipolar disorder: a randomized double-blind placebo-controlled trial. Acta Med Iran 2014; 52(10): 734-9.
[PMID: 25369006]

[244]   Faghihi T, Jahed A, Mahmoudi-Gharaei J, Sharifi V, Akhondzadeh S, Ghaeli P. Role of Omega-3 fatty acids in preventing metabolic disturbances in patients on olanzapine plus either sodium valproate or lithium: a randomized double-blind placebo-controlled trial. Daru 2012; 20(1): 43.
[http://dx.doi.org/10.1186/2008-2231-20-43] [PMID: 23351198]

[245]   Huang M, Yu L, Pan F, *et al.* A randomized, 13-week study assessing the efficacy and metabolic effects of paliperidone palmitate injection and olanzapine in first-episode schizophrenia patients. Prog Neuropsychopharmacol Biol Psychiatry 2018; 81: 122-30.
[http://dx.doi.org/10.1016/j.pnpbp.2017.10.021] [PMID: 29097257]

[246]   Hu S, Yao M, Peterson BS, *et al.* A randomized, 12-week study of the effects of extended-release paliperidone (paliperidone ER) and olanzapine on metabolic profile, weight, insulin resistance, and β-cell function in schizophrenic patients. Psychopharmacology (Berl) 2013; 230(1): 3-13.
[http://dx.doi.org/10.1007/s00213-013-3073-1] [PMID: 23559220]

[247]   Schreiner A, Niehaus D, Shuriquie NA, *et al.* Metabolic effects of paliperidone extended release *versus* oral olanzapine in patients with schizophrenia: a prospective, randomized, controlled trial. J Clin Psychopharmacol 2012; 32(4): 449-57.
[http://dx.doi.org/10.1097/JCP.0b013e31825cccad] [PMID: 22722501]

[248]   Dehelean L, Andor M, Romoşan AM, *et al.* Pharmacological and disorder associated cardiovascular changes in patients with psychosis. A comparison between olanzapine and risperidone. Farmacia 2018; 66(1): 129-34.

[249]   Kaushal J, Bhutani G, Gupta R. Comparison of fasting blood sugar and serum lipid profile changes after treatment with atypical antipsychotics olanzapine and risperidone. Singapore Med J 2012; 53(7): 488-92.
[PMID: 22815019]

[250]   Smith RC, Lindenmayer J-P, Hu Q, *et al.* Effects of olanzapine and risperidone on lipid metabolism in chronic schizophrenic patients with long-term antipsychotic treatment: a randomized five month study. Schizophr Res 2010; 120(1-3): 204-9.
[http://dx.doi.org/10.1016/j.schres.2010.04.001] [PMID: 20457512]

[251]   Wang H-H, Cai M, Wang H-N, *et al.* An assessor-blinded, randomized comparison of efficacy and tolerability of switching from olanzapine to ziprasidone and the combination of both in schizophrenia spectrum disorders. J Psychiatr Res 2017; 85: 59-65.
[http://dx.doi.org/10.1016/j.jpsychires.2016.11.002] [PMID: 27837658]

[252]   Ou J-J, Xu Y, Chen H-H, *et al.* Comparison of metabolic effects of ziprasidone *versus* olanzapine treatment in patients with first-episode schizophrenia. Psychopharmacology (Berl) 2013; 225(3): 627-35.
[http://dx.doi.org/10.1007/s00213-012-2850-6] [PMID: 22926006]

[253]   Grootens KP, van Veelen NMJ, Peuskens J, *et al.* Ziprasidone *vs.* olanzapine in recent-onset schizophrenia and schizoaffective disorder: results of an 8-week double-blind randomized controlled

trial. Schizophr Bull 2011; 37(2): 352-61.
[http://dx.doi.org/10.1093/schbul/sbp037] [PMID: 19542525]

[254] Ozguven HD, Baskak B, Oner O, Atbasoglu C. Metabolic effects of olanzapine and quetiapine: A six-week randomized, single blind, controlled study. Open Neuropsychopharmacol J 2011; 4(1): 10-7.
[http://dx.doi.org/10.2174/1876523801104010010]

[255] Lee H-B, Yoon B-H, Kwon Y-J, *et al.* The efficacy and safety of switching to ziprasidone from olanzapine in patients with bipolar I disorder: an 8-week, multicenter, open-label study. Clin Drug Investig 2013; 33(10): 743-53.
[http://dx.doi.org/10.1007/s40261-013-0120-y] [PMID: 23990283]

[256] Meyer JM, Pandina G, Bossie CA, Turkoz I, Greenspan A. Effects of switching from olanzapine to risperidone on the prevalence of the metabolic syndrome in overweight or obese patients with schizophrenia or schizoaffective disorder: analysis of a multicenter, rater-blinded, open-label study. Clin Ther 2005; 27(12): 1930-41.
[http://dx.doi.org/10.1016/j.clinthera.2005.12.005] [PMID: 16507379]

[257] Gupta S, Masand PS, Virk S, *et al.* Weight decline in patients switching from olanzapine to quetiapine. Schizophr Res 2004; 70(1): 57-62.
[http://dx.doi.org/10.1016/j.schres.2003.09.016] [PMID: 15246464]

[258] Bogusz MJ, Krüger KD, Maier RD, Erkwoh R, Tuchtenhagen F. Monitoring of olanzapine in serum by liquid chromatography-atmospheric pressure chemical ionization mass spectrometry. J Chromatogr B Biomed Sci Appl 1999; 732(2): 257-69.
[http://dx.doi.org/10.1016/S0378-4347(99)00287-X] [PMID: 10517347]

[259] Catlow JT, Barton RD, Clemens M, Gillespie TA, Goodwin M, Swanson SP. Analysis of olanzapine in human plasma utilizing reversed-phase high-performance liquid chromatography with electrochemical detection. J Chromatogr B Biomed Appl 1995; 668(1): 85-90.
[http://dx.doi.org/10.1016/0378-4347(95)00061-M] [PMID: 7550985]

[260] Kasper SC, Mattiuz EL, Swanson SP, Chiu JA, Johnson JT, Garner CO. Determination of olanzapine in human breast milk by high-performance liquid chromatography with electrochemical detection. J Chromatogr B Biomed Sci Appl 1999; 726(1-2): 203-9.
[http://dx.doi.org/10.1016/S0378-4347(99)00017-1] [PMID: 10348187]

[261] Aravagiri M, Ames D, Wirshing WC, Marder SR. Plasma level monitoring of olanzapine in patients with schizophrenia: determination by high-performance liquid chromatography with electrochemical detection. Ther Drug Monit 1997; 19(3): 307-13.
[http://dx.doi.org/10.1097/00007691-199706000-00011] [PMID: 9200772]

[262] Jenkins AJ, Sarconi KM, Raaf HN. Determination of olanzapine in a postmortem case. J Anal Toxicol 1998; 22(7): 605-9.
[http://dx.doi.org/10.1093/jat/22.7.605] [PMID: 9847013]

[263] Berna M, Shugert R, Mullen J. Determination of olanzapine in human plasma and serum by liquid chromatography/tandem mass spectrometry. J Mass Spectrom 1998; 33(10): 1003-8.
[http://dx.doi.org/10.1002/(SICI)1096-9888(1998100)33:10<1003::AID-JMS716>3.0.CO;2-P]
[PMID: 9821331]

[264] Olesen OV, Poulsen B, Linnet K. Fully automated on-line determination of olanzapine in serum for routine therapeutic drug monitoring. Ther Drug Monit 2001; 23(1): 51-5.
[http://dx.doi.org/10.1097/00007691-200102000-00010] [PMID: 11206044]

[265] Raggi MA, Casamenti G, Mandrioli R, Volterra V. A sensitive high-performance liquid chromatographic method using electrochemical detection for the analysis of olanzapine and desmethylolanzapine in plasma of schizophrenic patients using a new solid-phase extraction procedure. J Chromatogr B Biomed Sci Appl 2001; 750(1): 137-46.
[http://dx.doi.org/10.1016/S0378-4347(00)00438-2] [PMID: 11204214]

[266] Raggi MA, Mandrioli R, Sabbioni C, Ghedini N, Fanali S, Volterra V. Determination of olanzapine

and desmethylolanzapine in plasma of schizophrenic patients by an improved HPLC method with amperometric detection. Chromatographia 2001; 54: 203-7.
[http://dx.doi.org/10.1007/BF02492246]

[267]   Raggi MA, Casamenti G, Mandrioli R, Fanali S, De Ronchi D, Volterra V. Determination of the novel antipsychotic drug olanzapine in human plasma using HPLC with amperometric detection. Chromatographia 2000; 51: 562-6.
[http://dx.doi.org/10.1007/BF02490813]

[268]   Boulton DW, Markowitz JS, DeVane CL. A high-performance liquid chromatography assay with ultraviolet detection for olanzapine in human plasma and urine. J Chromatogr B Biomed Sci Appl 2001; 759(2): 319-23.
[http://dx.doi.org/10.1016/S0378-4347(01)00240-7] [PMID: 11499485]

[269]   Dusci LJ, Peter Hackett L, Fellows LM, Ilett KF. Determination of olanzapine in plasma by high-performance liquid chromatography using ultraviolet absorbance detection. J Chromatogr B Analyt Technol Biomed Life Sci 2002; 773(2): 191-7.
[http://dx.doi.org/10.1016/S1570-0232(02)00164-2] [PMID: 12031846]

[270]   Llorca PM, Coudore F, Corpelet C, Buyens A, Hoareau M, Eschalier A. Integration of olanzapine determinations in a HPLC-diode array detection system for routine psychotropic drug monitoring (2001). Clin Chem 2001; 47: 1719-21.

[271]   Siva Selva Kumar M, Ramanathan M. Concurrent determination of olanzapine, risperidone and 9-hydroxyrisperidone in human plasma by ultra performance liquid chromatography with diode array detection method: application to pharmacokinetic study. Biomed Chromatogr 2016; 30(2): 263-8.
[http://dx.doi.org/10.1002/bmc.3545] [PMID: 26129833]

[272]   da Fonseca BM, Moreno IED, Barroso M, Costa S, Queiroz JA, Gallardo E. Determination of seven selected antipsychotic drugs in human plasma using microextraction in packed sorbent and gas chromatography-tandem mass spectrometry. Anal Bioanal Chem 2013; 405(12): 3953-63.
[http://dx.doi.org/10.1007/s00216-012-6695-y] [PMID: 23314486]

[273]   D'Arrigo C, Migliardi G, Santoro V, Spina E. Determination of olanzapine in human plasma by reversed-phase high-performance liquid chromatography with ultraviolet detection. Ther Drug Monit 2006; 28(3): 388-93.
[http://dx.doi.org/10.1097/01.ftd.0000211800.66569.c9] [PMID: 16778724]

[274]   Ulrich S. Assay of olanzapine in human plasma by a rapid and sensitive gas chromatography-nitrogen phosphorus selective detection (GC-NPD) method: validation and comparison with high-performance liquid chromatography-coulometric detection. Ther Drug Monit 2005; 27(4): 463-8.
[http://dx.doi.org/10.1097/01.ftd.0000164245.24358.15] [PMID: 16044103]

[275]   Sabbioni C, Saracino MA, Mandrioli R, Albers L, Boncompagni G, Raggi MA. Rapid analysis of olanzapine and desmethylolanzapine in human plasma using high-performance liquid chromatography with coulometric detection. Anal Chim Acta 2004; 516(1-2): 111-7.
[http://dx.doi.org/10.1016/j.aca.2004.04.031]

[276]   Sachse J, Köller J, Härtter S, Hiemke C. Automated analysis of quetiapine and other antipsychotic drugs in human blood by high performance-liquid chromatography with column-switching and spectrophotometric detection. J Chromatogr B Analyt Technol Biomed Life Sci 2006; 830(2): 342-8.
[http://dx.doi.org/10.1016/j.jchromb.2005.11.032] [PMID: 16337441]

[277]   Berna M, Ackermann B, Ruterbories K, Glass S. Determination of olanzapine in human blood by liquid chromatography-tandem mass spectrometry. J Chromatogr B Analyt Technol Biomed Life Sci 2002; 767(1): 163-8.
[http://dx.doi.org/10.1016/S0378-4347(01)00548-5] [PMID: 11863288]

[278]   Ni X-J, Wang Z-Z, Shang D-W, Lu H-Y, Zhang M, Wen Y-G. Simultaneous analysis of olanzapine, fluoxetine, and norfluoxetine in human plasma using liquid chromatography-mass spectrometry and its application to a pharmacokinetic study. J Chromatogr B Analyt Technol Biomed Life Sci 2018; 1092:

506-14.
[http://dx.doi.org/10.1016/j.jchromb.2018.05.026] [PMID: 30008307]

[279]   Ravinder S, Bapuji AT, Mukkanti K, Reddy DC. Simultaneous determination of olanzapine and fluoxetine in human plasma by LC-MS/MS and its application to pharmacokinetic study. J Liq Chromatogr Relat Technol 2013; 36(19): 2651-68.
[http://dx.doi.org/10.1080/10826076.2012.723098]

[280]   Gopinath S, Kumar RS, Alexander S, Danabal P. Development of a rapid and sensitive SPE-L--MS/MS method for the simultaneous estimation of fluoxetine and olanzapine in human plasma. Biomed Chromatogr 2012; 26(9): 1077-82.
[http://dx.doi.org/10.1002/bmc.1750] [PMID: 22113919]

[281]   Josefsson M, Roman M, Skogh E, Dahl M-L. Liquid chromatography/tandem mass spectrometry method for determination of olanzapine and N-desmethylolanzapine in human serum and cerebrospinal fluid. J Pharm Biomed Anal 2010; 53(3): 576-82.
[http://dx.doi.org/10.1016/j.jpba.2010.03.040] [PMID: 20452161]

[282]   Nirogi RVS, Kandikere VN, Shukla M, *et al.* Development and validation of a sensitive liquid chromatography/electrospray tandem mass spectrometry assay for the quantification of olanzapine in human plasma. J Pharm Biomed Anal 2006; 41(3): 935-42.
[http://dx.doi.org/10.1016/j.jpba.2006.01.040] [PMID: 16504450]

[283]   Zhou Z, Li X, Li K, *et al.* Simultaneous determination of clozapine, olanzapine, risperidone and quetiapine in plasma by high-performance liquid chromatography-electrospray ionization mass spectrometry. J Chromatogr B Analyt Technol Biomed Life Sci 2004; 802(2): 257-62.
[http://dx.doi.org/10.1016/j.jchromb.2003.11.037] [PMID: 15018785]

[284]   Pinto MAL, de Souza ID, Queiroz MEC. Determination of drugs in plasma samples by disposable pipette extraction with C18-BSA phase and liquid chromatography-tandem mass spectrometry. J Pharm Biomed Anal 2017; 139: 116-24.
[http://dx.doi.org/10.1016/j.jpba.2017.02.052] [PMID: 28279926]

[285]   El-Saifi N, Moyle W, Jones C, Tuffaha H. Quetiapine safety in older adults: a systematic literature review. J Clin Pharm Ther 2016; 41(1): 7-18.
[http://dx.doi.org/10.1111/jcpt.12357] [PMID: 26813985]

[286]   Marlowe KF, Howard D, Chung A. New onset diabetes with ketoacidosis attributed to quetiapine. South Med J 2007; 100(8): 829-31.
[http://dx.doi.org/10.1097/SMJ.0b013e31804b1e4d] [PMID: 17713311]

[287]   Carnahan RM, Reantaso AA, Teegarden BA, Pogue T. Severe hyperlipidemia associated with olanzapine and quetiapine use. Am J Psychiatry 2007; 164(10): 1614-5.
[http://dx.doi.org/10.1176/appi.ajp.2007.07020253] [PMID: 17898359]

[288]   Jalota R, Bond C, José RJ. Quetiapine and the development of the metabolic syndrome. QJM 2015; 108(3): 245-7.
[http://dx.doi.org/10.1093/qjmed/hcs142] [PMID: 22908317]

[289]   Madsen KR. Fatal hypertriglyceridaemia, acute pancreatitis and diabetic ketoacidosis possibly induced by quetiapine. BMJ Case Reports 2014; bcr2013202039.
[http://dx.doi.org/10.1136/bcr-2013-202039]

[290]   Mancano MA. Pancreatitis-associated with riluzole; linezolid-induced hypoglycemia; sorafenib-induced acute generalized exanthematous pustulosis; creatine supplementation-induced thrombotic events; acute pancreatitis associated with quetiapine; hypomagnesemia and seizure associated with rabeprazole. Hosp Pharm 2014; 49(11): 1004-8.
[http://dx.doi.org/10.1310/hpj4911-1004] [PMID: 25673887]

[291]   Chen C-H, Lin T-Y, Chen T-T, *et al.* A prospective study of glucose homeostasis in quetiapine-treated schizophrenic patients by using the intravenous glucose tolerance test. Prog Neuropsychopharmacol Biol Psychiatry 2011; 35(4): 965-9.

[http://dx.doi.org/10.1016/j.pnpbp.2011.01.015] [PMID: 21291941]

[292] Koller EA, Weber J, Doraiswamy PM, Schneider BS. A survey of reports of quetiapine-associated hyperglycemia and diabetes mellitus. J Clin Psychiatry 2004; 65(6): 857-63.
[http://dx.doi.org/10.4088/JCP.v65n0619] [PMID: 15291665]

[293] Atmaca M, Kuloglu M, Tezcan E, Ustundag B. Serum leptin and triglyceride levels in patients on treatment with atypical antipsychotics. J Clin Psychiatry 2003; 64(5): 598-604.
[http://dx.doi.org/10.4088/JCP.v64n0516] [PMID: 12755665]

[294] Meyer JM, Koro CE. The effects of antipsychotic therapy on serum lipids: a comprehensive review. Schizophr Res 2004; 70(1): 1-17.
[http://dx.doi.org/10.1016/j.schres.2004.01.014] [PMID: 15246458]

[295] Meltzer HY. Focus on the metabolic consequences of long-term treatment with olanzapine, quetiapine and risperidone: are there differences? Int J Neuropsychopharmacol 2005; 8(2): 153-6.
[http://dx.doi.org/10.1017/S1461145705005183] [PMID: 15780147]

[296] Asmal L, Flegar SJ, Wang J, Rummel-Kluge C, Komossa K, Leucht S. Quetiapine *versus* other atypical antipsychotics for schizophrenia. Cochrane Database Syst Rev 2013; (11): CD006625
[http://dx.doi.org/10.1002/14651858.CD006625.pub3] [PMID: 24249315]

[297] Karayal ON, Glue P, Bachinsky M, *et al.* Switching from quetiapine to ziprasidone: a sixteen-week, open-label, multicenter study evaluating the effectiveness and safety of ziprasidone in outpatient subjects with schizophrenia or schizoaffective disorder. J Psychiatr Pract 2011; 17(2): 100-9.
[http://dx.doi.org/10.1097/01.pra.0000396061.05269.c8] [PMID: 21430488]

[298] Liang C-S, Yang F-W, Lo S-M. Rapid development of severe hypertriglyceridemia and hypercholesterolemia during augmentation of quetiapine with valproic acid. J Clin Psychopharmacol 2011; 31(2): 242-3.
[http://dx.doi.org/10.1097/JCP.0b013e31820f4f9e] [PMID: 21364335]

[299] Davis PC, Wong J, Gefvert O. Analysis and pharmacokinetics of quetiapine and two metabolites in human plasma using reversed-phase HPLC with ultraviolet and electrochemical detection. J Pharm Biomed Anal 1999; 20(1-2): 271-82.
[http://dx.doi.org/10.1016/S0731-7085(99)00036-9] [PMID: 10704032]

[300] Prakash D, Bhat K, Shetty R, Chaudhary P, Pathak SR, Ghosh A. Quantification of quetiapine in human plasma by reverse phase high performance liquid chromatography. Arzneimittel-Forschung/Drug Research 2010; 60(11): 654-9.

[301] Mandrioli R, Fanali S, Ferranti A, Raggi MA. HPLC analysis of the novel antipsychotic drug quetiapine in human plasma. J Pharm Biomed Anal 2002; 30(4): 969-77.
[http://dx.doi.org/10.1016/S0731-7085(02)00395-3] [PMID: 12408887]

[302] Wróblewski K, Petruczynik A, Waksmundzka-Hajnos M. Separation and determination of selected psychotropic drugs in human serum by SPE/HPLC/DAD on C18 and Polar-RP columns. J Liq Chromatogr Relat Technol 2017; 40(2): 75-82.
[http://dx.doi.org/10.1080/10826076.2017.1284675]

[303] Li D, Zou J, Cai P-S, Xiong C-M, Ruan J-L. Preparation of magnetic ODS-PAN thin-films for microextraction of quetiapine and clozapine in plasma and urine samples followed by HPLC-UV detection. J Pharm Biomed Anal 2016; 125: 319-28.
[http://dx.doi.org/10.1016/j.jpba.2016.04.006] [PMID: 27085135]

[304] Nirogi R, Bhyrapuneni G, Kandikere V, Mudigonda K, Ajjala D, Mukkanti K. Sensitive liquid chromatography tandem mass spectrometry method for the quantification of Quetiapine in plasma. Biomed Chromatogr 2008; 22(10): 1043-55.
[http://dx.doi.org/10.1002/bmc.1012] [PMID: 18781706]

[305] Barrett B, Holčapek M, Huclová J, *et al.* Validated HPLC-MS/MS method for determination of quetiapine in human plasma. J Pharm Biomed Anal 2007; 44(2): 498-505.

[http://dx.doi.org/10.1016/j.jpba.2007.03.034] [PMID: 17499470]

[306] Saracino MA, Mercolini L, Flotta G, Albers LJ, Merli R, Raggi MA. Simultaneous determination of fluvoxamine isomers and quetiapine in human plasma by means of high-performance liquid chromatography. J Chromatogr B Analyt Technol Biomed Life Sci 2006; 843(2): 227-33.
[http://dx.doi.org/10.1016/j.jchromb.2006.06.001] [PMID: 16798118]

[307] Li M, Zhang S, Shi A, Qi W, Liu Y. Determination of quetiapine in human plasma by LC-MS/MS and its application in a bioequivalence study. J Chromatogr B Analyt Technol Biomed Life Sci 2017; 1060: 10-4.
[http://dx.doi.org/10.1016/j.jchromb.2017.05.031] [PMID: 28578191]

[308] Xiong X, Yang L, Duan J. Development and validation of a sensitive and robust LC-MS/MS with electrospray ionization method for simultaneous quantitation of quetiapine and its active metabolite norquetiapine in human plasma. Clin Chim Acta 2013; 423: 69-74.
[http://dx.doi.org/10.1016/j.cca.2013.04.016] [PMID: 23623924]

[309] Davis PC, Bravo O, Gehrke M, Azumaya CT. Development and validation of an LC-MS/MS method for the determination of quetiapine and four related metabolites in human plasma. J Pharm Biomed Anal 2010; 51(5): 1113-9.
[http://dx.doi.org/10.1016/j.jpba.2009.11.018] [PMID: 20022726]

[310] Scahill L, Jeon S, Boorin SJ, *et al.* Weight Gain and Metabolic Consequences of Risperidone in Young Children With Autism Spectrum Disorder. J Am Acad Child Adolesc Psychiatry 2016; 55(5): 415-23.
[http://dx.doi.org/10.1016/j.jaac.2016.02.016] [PMID: 27126856]

[311] Matera E, Margari L, Palmieri VO, Zagaria G, Palumbi R, Margari F. Risperidone and cardiometabolic risk in children and adolescents: Clinical and instrumental issues. J Clin Psychopharmacol 2017; 37(3): 302-9.
[http://dx.doi.org/10.1097/JCP.0000000000000688] [PMID: 28338545]

[312] Goeb J-L, Marco S, Duhamel A, *et al.* [Metabolic side effects of risperidone in early onset schizophrenia]. Encephale 2010; 36(3): 242-52.
[http://dx.doi.org/10.1016/j.encep.2009.10.008] [PMID: 20620267]

[313] Calarge CA, Acion L, Kuperman S, Tansey M, Schlechte JA. Weight gain and metabolic abnormalities during extended risperidone treatment in children and adolescents. J Child Adolesc Psychopharmacol 2009; 19(2): 101-9.
[http://dx.doi.org/10.1089/cap.2008.007] [PMID: 19364288]

[314] Sahoo S, Mishra B, Akhtar S. Dose-dependent acute excessive weight gain and metabolic changes in a drug-naive patient on risperidone are reversible with discontinuation: a case report. Br J Clin Pharmacol 2007; 64(5): 715-6.
[http://dx.doi.org/10.1111/j.1365-2125.2007.02941.x] [PMID: 17509037]

[315] Calarge CA, Nicol G, Schlechte JA, Burns TL. Cardiometabolic outcomes in children and adolescents following discontinuation of long-term risperidone treatment. J Child Adolesc Psychopharmacol 2014; 24(3): 120-9.
[http://dx.doi.org/10.1089/cap.2013.0126] [PMID: 24725198]

[316] Németh G, Laszlovszky I, Czobor P, *et al.* Cariprazine *versus* risperidone monotherapy for treatment of predominant negative symptoms in patients with schizophrenia: a randomised, double-blind, controlled trial. Lancet 2017; 389(10074): 1103-13.
[http://dx.doi.org/10.1016/S0140-6736(17)30060-0] [PMID: 28185672]

[317] Khan RA, Mican LM, Suehs BT. Effects of olanzapine and risperidone on metabolic factors in children and adolescents: a retrospective evaluation. J Psychiatr Pract 2009; 15(4): 320-8.
[http://dx.doi.org/10.1097/01.pra.0000358319.81307.a5] [PMID: 19625888]

[318] Fu D-J, Bossie CA, Sliwa JK, Ma Y-W, Alphs L. Paliperidone palmitate *versus* oral risperidone and risperidone long-acting injection in patients with recently diagnosed schizophrenia: a tolerability and efficacy comparison. Int Clin Psychopharmacol 2014; 29(1): 45-55.

[http://dx.doi.org/10.1097/YIC.0000000000000006] [PMID: 24113628]

[319] Kim S-W, Chung Y-C, Lee Y-H, *et al.* Paliperidone ER *versus* risperidone for neurocognitive function in patients with schizophrenia: a randomized, open-label, controlled trial. Int Clin Psychopharmacol 2012; 27(5): 267-74.
[http://dx.doi.org/10.1097/YIC.0b013e328356acad] [PMID: 22809972]

[320] De Hert M, Mittoux A, He Y, Peuskens J. Metabolic parameters in the short- and long-term treatment of schizophrenia with sertindole or risperidone. Eur Arch Psychiatry Clin Neurosci 2011; 261(4): 231-9.
[http://dx.doi.org/10.1007/s00406-010-0142-x] [PMID: 20820795]

[321] Addington DE, Labelle A, Kulkarni J, Johnson G, Loebel A, Mandel FS. A comparison of ziprasidone and risperidone in the long-term treatment of schizophrenia: a 44-week, double-blind, continuation study. Can J Psychiatry 2009; 54(1): 46-54.
[http://dx.doi.org/10.1177/070674370905400108] [PMID: 19175979]

[322] Sönmez B, Vardar E, Altun GD, Abay E, Bedel D. Ziprasidone *versus* risperidone: Comparison of clinical efficacy and cardiac, extrapyramidal, and metabolic side effects in patients with acute exacerbation of schizophrenia and Schizoaffective disorders. Klinik Psikofarmakol Bülteni 2009; 19(2): 101-12.

[323] Lin C-H, Kuo C-C, Chou L-S, *et al.* A randomized, double-blind comparison of risperidone *versus* low-dose risperidone plus low-dose haloperidol in treating schizophrenia. J Clin Psychopharmacol 2010; 30(5): 518-25.
[http://dx.doi.org/10.1097/JCP.0b013e3181f28dff] [PMID: 20814315]

[324] Avenoso A, Facciolà G, Salemi M, Spina E. Determination of risperidone and its major metabolite 9-hydroxyrisperidone in human plasma by reversed-phase liquid chromatography with ultraviolet detection. J Chromatogr B Biomed Sci Appl 2000; 746(2): 173-81.
[http://dx.doi.org/10.1016/S0378-4347(00)00323-6] [PMID: 11076069]

[325] Nagasaki T, Ohkubo T, Sugawara K, Yasui N, Furukori H, Kaneko S. Determination of risperidone and 9-hydroxyrisperidone in human plasma by high-performance liquid chromatography: application to therapeutic drug monitoring in Japanese patients with schizophrenia. J Pharm Biomed Anal 1999; 19(3-4): 595-601.
[http://dx.doi.org/10.1016/S0731-7085(98)00261-1] [PMID: 10704125]

[326] Jones T, Van Breda K, Charles B, Dean AJ, McDermott BM, Norris R. Determination of risperidone and 9-Hydroxyrisperidone using HPLC, in plasma of children and adolescents with emotional and behavioural disorders. Biomed Chromatogr 2009; 23(9): 929-34.
[http://dx.doi.org/10.1002/bmc.1204] [PMID: 19353731]

[327] Mercolini L, Grillo M, Bartoletti C, Boncompagni G, Raggi MA. Simultaneous analysis of classical neuroleptics, atypical antipsychotics and their metabolites in human plasma. Anal Bioanal Chem 2007; 388(1): 235-43.
[http://dx.doi.org/10.1007/s00216-007-1195-1] [PMID: 17340084]

[328] Balant-Gorgia AE, Gex-Fabry M, Genet C, Balant LP. Therapeutic drug monitoring of risperidone using a new, rapid HPLC method: reappraisal of interindividual variability factors. Ther Drug Monit 1999; 21(1): 105-15.
[http://dx.doi.org/10.1097/00007691-199902000-00017] [PMID: 10051063]

[329] Aravagiri M, Marder SR, Van Putten T, Midha KK. Determination of risperidone in plasma by high-performance liquid chromatography with electrochemical detection: application to therapeutic drug monitoring in schizophrenic patients. J Pharm Sci 1993; 82(5): 447-9.
[http://dx.doi.org/10.1002/jps.2600820503] [PMID: 7689651]

[330] Aravagiri M, Marder SR, Wirshing D, Wirshing WC. Plasma concentrations of risperidone and its 9-hydroxy metabolite and their relationship to dose in schizophrenic patients: simultaneous determination by a high performance liquid chromatography with electrochemical detection.

Pharmacopsychiatry 1998; 31(3): 102-9.
[http://dx.doi.org/10.1055/s-2007-979308] [PMID: 9657237]

[331] Le Moing JP, Edouard S, Levron JC. Determination of risperidone and 9-hydroxyrisperidone in human plasma by high-performance liquid chromatography with electrochemical detection. J Chromatogr A 1993; 614(2): 333-9.
[http://dx.doi.org/10.1016/0378-4347(93)80327-Z] [PMID: 7686177]

[332] Mandrioli R, Mercolini L, Lateana D, Boncompagni G, Raggi MA. Analysis of risperidone and 9-hydroxyrisperidone in human plasma, urine and saliva by MEPS-LC-UV. J Chromatogr B Analyt Technol Biomed Life Sci 2011; 879(2): 167-73.
[http://dx.doi.org/10.1016/j.jchromb.2010.11.033] [PMID: 21183412]

[333] Schatz DS, Saria A. Simultaneous determination of paroxetine, risperidone and 9-hydroxyrisperidone in human plasma by high-performance liquid chromatography with coulometric detection. Pharmacology 2000; 60(1): 51-6.
[http://dx.doi.org/10.1159/000028347] [PMID: 10629444]

[334] Price MC, Hoffman DW. Therapeutic drug monitoring of risperidone and 9-hydroxyrisperidone in serum with solid-phase extraction and high-performance liquid chromatography. Ther Drug Monit 1997; 19(3): 333-7.
[http://dx.doi.org/10.1097/00007691-199706000-00015] [PMID: 9200776]

[335] Aravagiri M, Marder SR. Simultaneous determination of risperidone and 9-hydroxyrisperidone in plasma by liquid chromatography/electrospray tandem mass spectrometry. J Mass Spectrom 2000; 35(6): 718-24.
[http://dx.doi.org/10.1002/1096-9888(200006)35:6<718::AID-JMS999>3.0.CO;2-O]   [PMID: 10862124]

[336] Vanwong N, Prommas S, Puangpetch A, *et al.* Development and Validation of Liquid Chromatography/Tandem Mass Spectrometry Analysis for Therapeutic Drug Monitoring of Risperidone and 9-Hydroxyrisperidone in Pediatric Patients with Autism Spectrum Disorders. J Clin Lab Anal 2016; 30(6): 1236-46.
[http://dx.doi.org/10.1002/jcla.22009] [PMID: 27346210]

[337] Zhang X, Zhao X, Zhang C, *et al.* Accuracy profile theory for the validation of an LC-MS-MS Method for the Determination of Risperidone and 9-Hydroxyrisperidone in Human Plasma. Chromatographia 2010; 71(11-12): 1015-23.
[http://dx.doi.org/10.1365/s10337-010-1580-3]

[338] Bhatt J, Subbaiah G, Singh S. Liquid chromatography/tandem mass spectrometry method for simultaneous determination of risperidone and its active metabolite 9-hydroxyrisperidone in human plasma. Rapid Commun Mass Spectrom 2006; 20(14): 2109-14.
[http://dx.doi.org/10.1002/rcm.2537] [PMID: 16775814]

[339] McClean S, O'Kane EJ, Smyth WF. Electrospray ionisation-mass spectrometric characterisation of selected anti-psychotic drugs and their detection and determination in human hair samples by liquid chromatography-tandem mass spectrometry. J Chromatogr B Biomed Sci Appl 2000; 740(2): 141-57.
[http://dx.doi.org/10.1016/S0378-4347(00)00038-4] [PMID: 10821400]

[340] Flarakos J, Luo W, Aman M, Svinarov D, Gerber N, Vouros P. Quantification of risperidone and 9-hydroxyrisperidone in plasma and saliva from adult and pediatric patients by liquid chromatography-mass spectrometry. J Chromatogr A 2004; 1026(1-2): 175-83.
[http://dx.doi.org/10.1016/j.chroma.2003.10.138] [PMID: 14763744]

[341] Belotto KCR, Raposo NRB, Ferreira AS, Gattaz WF. Relative bioavailability of two oral formulations of risperidone 2 mg: A single-dose, randomized-sequence, open-label, two-period crossover comparison in healthy Brazilian volunteers. Clin Ther 2010; 32(12): 2106-15.
[http://dx.doi.org/10.1016/j.clinthera.2010.11.006] [PMID: 21118746]

[342] Moody DE, Laycock JD, Huang W, Foltz RL. A high-performance liquid chromatographic-

atmospheric pressure chemical ionization-tandem mass spectrometric method for determination of risperidone and 9-hydroxyrisperidone in human plasma. J Anal Toxicol 2004; 28(6): 494-7.
[http://dx.doi.org/10.1093/jat/28.6.494] [PMID: 15516302]

[343]   Locatelli I, Mrhar A, Grabnar I. Simultaneous determination of risperidone and 9-hydroxyrisperidone enantiomers in human blood plasma by liquid chromatography with electrochemical detection. J Pharm Biomed Anal 2009; 50(5): 905-10.
[http://dx.doi.org/10.1016/j.jpba.2009.06.013] [PMID: 19589654]

[344]   De Meulder M, Remmerie BMM, de Vries R, *et al.* Validated LC-MS/MS methods for the determination of risperidone and the enantiomers of 9-hydroxyrisperidone in human plasma and urine. J Chromatogr B Analyt Technol Biomed Life Sci 2008; 870(1): 8-16.
[http://dx.doi.org/10.1016/j.jchromb.2008.04.041] [PMID: 18571483]

[345]   Čabovska B, Cox SL, Vinks AA. Determination of risperidone and enantiomers of 9-hydroxyrisperidone in plasma by LC-MS/MS. J Chromatogr B Analyt Technol Biomed Life Sci 2007; 852(1-2): 497-504.
[http://dx.doi.org/10.1016/j.jchromb.2007.02.007] [PMID: 17344104]

[346]   European Agency for the Evaluation of Medicinal Products. Agency for the Evaluation of Medicinal Products. Committee for proprietary medicinal products opinion following an article 36 referral: sertindole 2002EMA, London Available from: https://wwwemaeuropaeu/documents/referral/opinion-following-article-36-referral-sertindole-international-non-proprietary-name-inn-sertindole_enpdf , [last accessed on December 1, 2018];

[347]   Zoccali RA, Bruno A, Muscatello MRA. Efficacy and safety of sertindole in schizophrenia: a clinical review. J Clin Psychopharmacol 2015; 35(3): 286-95.
[http://dx.doi.org/10.1097/JCP.0000000000000305] [PMID: 25830594]

[348]   Kasper S. Sertindole: safety and tolerability profile. Int J Psychiatry Clin Pract 2002; 6(1): 27-32.
[http://dx.doi.org/10.1080/13651500215967] [PMID: 24931887]

[349]   Kane JM, Potkin SG, Daniel DG, Buckley PF. A double-blind, randomized study comparing the efficacy and safety of sertindole and risperidone in patients with treatment-resistant schizophrenia. J Clin Psychiatry 2011; 72(2): 194-204.
[http://dx.doi.org/10.4088/JCP.07m03733yel] [PMID: 20673553]

[350]   van Kammen DP, McEvoy JP, Targum SD, Kardatzke D, Sebree TB. A randomized, controlled, dose-ranging trial of sertindole in patients with schizophrenia. Psychopharmacology (Berl) 1996; 124(1-2): 168-75.
[http://dx.doi.org/10.1007/BF02245618] [PMID: 8935813]

[351]   Lewis R, Bagnall AM, Leitner M. Sertindole for schizophrenia. Cochrane Database Syst Rev 2005; 3(3)CD001715
[PMID: 16034864]

[352]   Daniel DG, Wozniak P, Mack RJ, McCarthy BG. Long-term efficacy and safety comparison of sertindole and haloperidol in the treatment of schizophrenia. Psychopharmacol Bull 1998; 34(1): 61-9.
[PMID: 9564200]

[353]   Tzeng T-B, Stamm G, Chu S-Y. Sensitive method for the assay of sertindole in plasma by high-performance liquid chromatography and fluorimetric detection. J Chromatogr B Biomed Appl 1994; 661(2): 299-306.
[http://dx.doi.org/10.1016/0378-4347(94)00356-4] [PMID: 7894670]

[354]   Canal-Raffin M, Déridet E, Titier K, Frakra E, Molimard M, Moore N. Simplified ultraviolet liquid chromatographic method for determination of sertindole, dehydrosertindole and norsertindole, in human plasma. J Chromatogr B Analyt Technol Biomed Life Sci 2005; 814(1): 61-7.
[http://dx.doi.org/10.1016/j.jchromb.2004.10.020] [PMID: 15607708]

[355]   Menacherry SD, Stamm GE, Chu S-Y. A sensitive and specific method for assay of sertindole and its metabolites in human, rat, dog, and mouse plasma using HPLC with tandem mass spectrometric

detection. J Liq Chromatogr Relat Technol 1997; 20: 2241-57.
[http://dx.doi.org/10.1080/10826079708006560]

[356] Altun Y, Dogan-Topal B, Uslu B, Ozkan SA. Anodic behavior of sertindole and its voltammetric determination in pharmaceuticals and human serum using glassy carbon and boron-doped diamond electrodes. Electrochim Acta 2009; 54(6): 1893-903.
[http://dx.doi.org/10.1016/j.electacta.2008.10.010]

[357] Findling RL, Cavuş I, Pappadopulos E, *et al*. Ziprasidone in adolescents with schizophrenia: results from a placebo-controlled efficacy and long-term open-extension study. J Child Adolesc Psychopharmacol 2013; 23(8): 531-44.
[http://dx.doi.org/10.1089/cap.2012.0068] [PMID: 24111983]

[358] Kemp DE, Karayal ON, Calabrese JR, *et al*. Ziprasidone with adjunctive mood stabilizer in the maintenance treatment of bipolar I disorder: long-term changes in weight and metabolic profiles. Eur Neuropsychopharmacol 2012; 22(2): 123-31.
[http://dx.doi.org/10.1016/j.euroneuro.2011.06.005] [PMID: 21798721]

[359] Vanderburg D, Keohane D, Karayal ON, Pappadopulos E. Ziprasidone and the relative risk of diabetes. Br J Psychiatry 2011; 198(2): 157-8.
[http://dx.doi.org/10.1192/bjp.198.2.157a] [PMID: 21282789]

[360] Meyer JM, Davis VG, Goff DC, *et al*. Change in metabolic syndrome parameters with antipsychotic treatment in the CATIE Schizophrenia Trial: prospective data from phase 1. Schizophr Res 2008; 101(1-3): 273-86.
[http://dx.doi.org/10.1016/j.schres.2007.12.487] [PMID: 18258416]

[361] Yood MU, DeLorenze G, Quesenberry CP Jr, *et al*. The incidence of diabetes in atypical antipsychotic users differs according to agent--results from a multisite epidemiologic study. Pharmacoepidemiol Drug Saf 2009; 18(9): 791-9.
[http://dx.doi.org/10.1002/pds.1781] [PMID: 19526626]

[362] Kahn R, Fleischhacker WW, Karayal O, Siu C, Pappadopulos E. EUFEST: the effects of first and second generation antipsychotics on metabolic and cardiovascular risk factors. American Psychiatric Association 162nd Annual Meeting.

[363] Simpson GM, Glick ID, Weiden PJ, Romano SJ, Siu CO. Randomized, controlled, double-blind multicenter comparison of the efficacy and tolerability of ziprasidone and olanzapine in acutely ill inpatients with schizophrenia or schizoaffective disorder. Am J Psychiatry 2004; 161(10): 1837-47.
[http://dx.doi.org/10.1176/ajp.161.10.1837] [PMID: 15465981]

[364] Kinon BJ, Lipkovich I, Edwards SB, Adams DH, Ascher-Svanum H, Siris SG. A 24-week randomized study of olanzapine *versus* ziprasidone in the treatment of schizophrenia or schizoaffective disorder in patients with prominent depressive symptoms. J Clin Psychopharmacol 2006; 26(2): 157-62.
[http://dx.doi.org/10.1097/01.jcp.0000204137.82298.06] [PMID: 16633144]

[365] Lee C-P, Chen AP-J, Juang Y-Y. Weight gain while switching from polypharmacy to ziprasidone: A case report. Clin Schizophr Relat Psychoses 2015; 9(3): 141-4.
[http://dx.doi.org/10.3371/CSRP.LECH.043013] [PMID: 23644168]

[366] Kessing LV, Thomsen AF, Mogensen UB, Andersen PK. Treatment with antipsychotics and the risk of diabetes in clinical practice. Br J Psychiatry 2010; 197(4): 266-71.
[http://dx.doi.org/10.1192/bjp.bp.109.076935] [PMID: 20884948]

[367] Mandrioli R, Protti M, Mercolini L. Evaluation of the pharmacokinetics, safety and clinical efficacy of ziprasidone for the treatment of schizophrenia and bipolar disorder. Expert Opin Drug Metab Toxicol 2015; 11(1): 149-74.
[http://dx.doi.org/10.1517/17425255.2015.991713] [PMID: 25483358]

[368] Chue P, Mandel FS, Therrien F. The effect of ziprasidone on metabolic syndrome risk factors in subjects with schizophrenia: a 1 year, open-label, prospective study. Curr Med Res Opin 2014; 30(6): 997-1005.

[http://dx.doi.org/10.1185/03007995.2014.898139] [PMID: 24568177]

[369] Rossi A, Vita A, Tiradritti P, Romeo F. Assessment of clinical and metabolic status, and subjective well-being, in schizophrenic patients switched from typical and atypical antipsychotics to ziprasidone. Int Clin Psychopharmacol 2008; 23(4): 216-22.
[http://dx.doi.org/10.1097/YIC.0b013e3282f94905] [PMID: 18545060]

[370] Montes JM, Rodriguez JL, Balbo E, *et al.* Improvement in antipsychotic-related metabolic disturbances in patients with schizophrenia switched to ziprasidone. Prog Neuropsychopharmacol Biol Psychiatry 2007; 31(2): 383-8.
[http://dx.doi.org/10.1016/j.pnpbp.2006.10.002] [PMID: 17129654]

[371] Lindenmayer J-P, Tedeschi F, Yusim A, *et al.* Ziprasidone's effect on metabolic markers in patients with diabetes and chronic schizophrenia. Clinical Schizophrenia and Related Psychoses 2012; 5(4): 185-92.
[http://dx.doi.org/10.3371/CSRP.5.4.2]

[372] Weiden PJ, Newcomer JW, Loebel AD, Yang R, Lebovitz HE. Long-term changes in weight and plasma lipids during maintenance treatment with ziprasidone. Neuropsychopharmacology 2008; 33(5): 985-94.
[http://dx.doi.org/10.1038/sj.npp.1301482] [PMID: 17637612]

[373] Janiszewski JS, Fouda HG, Cole RO. Development and validation of a high-sensitivity assay for an antipsychotic agent, CP-88,059, with solid-phase extraction and narrow-bore high-performance liquid chromatography. J Chromatogr B Biomed Appl 1995; 668(1): 133-9.
[http://dx.doi.org/10.1016/0378-4347(95)00071-P] [PMID: 7550970]

[374] Sachse J, Härtter S, Hiemke C. Automated determination of ziprasidone by HPLC with column switching and spectrophotometric detection. Ther Drug Monit 2005; 27(2): 158-62.
[http://dx.doi.org/10.1097/01.ftd.0000150879.36296.4d] [PMID: 15795645]

[375] Mercolini L, Protti M, Fulgenzi G, *et al.* A fast and feasible microextraction by packed sorbent (MEPS) procedure for HPLC analysis of the atypical antipsychotic ziprasidone in human plasma. J Pharm Biomed Anal 2014; 88: 467-71.
[http://dx.doi.org/10.1016/j.jpba.2013.09.019] [PMID: 24176751]

[376] Al-Dirbashi OY, Aboul-Enein HY, Al-Odaib A, Jacob M, Rashed MS. Rapid liquid chromatography-tandem mass spectrometry method for quantification of ziprasidone in human plasma. Biomed Chromatogr 2006; 20(4): 365-8.
[http://dx.doi.org/10.1002/bmc.571] [PMID: 16167302]

[377] Kirchherr H, Kühn-Velten WN. Quantitative determination of forty-eight antidepressants and antipsychotics in human serum by HPLC tandem mass spectrometry: a multi-level, single-sample approach. J Chromatogr B Analyt Technol Biomed Life Sci 2006; 843(1): 100-13.
[http://dx.doi.org/10.1016/j.jchromb.2006.05.031] [PMID: 16798119]

[378] Aravagiri M, Marder SR, Pollock B. Determination of ziprasidone in human plasma by liquid chromatography-electrospray tandem mass spectrometry and its application to plasma level determination in schizophrenia patients. J Chromatogr B Analyt Technol Biomed Life Sci 2007; 847(2): 237-44.
[http://dx.doi.org/10.1016/j.jchromb.2006.10.024] [PMID: 17098485]

[379] Hasselstrøm J. Quantification of antidepressants and antipsychotics in human serum by precipitation and ultra high pressure liquid chromatography-tandem mass spectrometry. J Chromatogr B Analyt Technol Biomed Life Sci 2011; 879(1): 123-8.
[http://dx.doi.org/10.1016/j.jchromb.2010.11.024] [PMID: 21163713]

[380] Gupta VK, Agarwal S, Singhal B. Potentiometric assay of antipsychotic drug (ziprasidone hydrochloride) in pharmaceuticals, serum and urine. Int J Electrochem Sci 2011; 6: 3036-56.

[381] García MS, Ortuño JA, Cuartero M, Abuherba MS. Use of a new ziprasidone-selective electrode in mixed solvents and its application in the analysis of pharmaceuticals and biological fluids. Sensors

(Basel) 2011; 11(9): 8813-25.
[http://dx.doi.org/10.3390/s110908813] [PMID: 22164107]

[382] Suckow RF, Fein M, Correll CU, Cooper TB. Determination of plasma ziprasidone using liquid chromatography with fluorescence detection. J Chromatogr B Analyt Technol Biomed Life Sci 2004; 799(2): 201-8.
[http://dx.doi.org/10.1016/j.jchromb.2003.10.027] [PMID: 14670738]

[383] Mercolini L, Mandrioli R, Protti M, Conca A, Albers LJ, Raggi MA. Dried blood spot testing: a novel approach for the therapeutic drug monitoring of ziprasidone-treated patients. Bioanalysis 2014; 6(11): 1487-95.
[http://dx.doi.org/10.4155/bio.14.3] [PMID: 25046049]

[384] Zhong Q, Shen L, Liu J, *et al.* Automatic on-line solid-phase extraction with ultra-high performance liquid chromatography and tandem mass spectrometry for the determination of ten antipsychotics in human plasma. J Sep Sci 2016; 39(11): 2129-37.
[http://dx.doi.org/10.1002/jssc.201600129] [PMID: 27060597]

CHAPTER 2

Insulin Therapy and Foetoplacental Endothelial Dysfunction in Gestational Diabetes Mellitus

Mario Subiabre[1,*], Roberto Villalobos-Labra[1], Luis Silva[1,2], Fabián Pardo[1,3] and Luis Sobrevia[1,4,5,*]

[1] *Cellular and Molecular Physiology Laboratory (CMPL), Department of Obstetrics, Division of Obstetrics and Gynaecology, School of Medicine, Faculty of Medicine, Pontificia Universidad Católica de Chile, Santiago 8330024, Chile*

[2] *Immunoendocrinology, Division of Medical Biology, Department of Pathology and Medical Biology, University of Groningen, University Medical Centre Groningen (UMCG), Groningen 9700 RB, The Netherlands*

[3] *Metabolic Diseases Research Laboratory, Center of Research, Development, and Innovation in Health - Aconcagua Valley, San Felipe Campus, School of Medicine, Faculty of Medicine, Universidad de Valparaíso, 2172972 San Felipe, Chile*

[4] *Department of Physiology, Faculty of Pharmacy, Universidad de Sevilla, Seville E-41012, Spain*

[5] *University of Queensland Centre for Clinical Research (UQCCR), Faculty of Medicine and Biomedical Sciences, University of Queensland, Herston, QLD 4029, Queensland, Australia*

Abstract: Gestational diabetes mellitus (GDM) is a condition characterised by glucose intolerance first diagnosed in pregnancy. The first line of treatment for women diagnosed with GDM is diet control (GDMd). However, some of these women even after diet persist continue showing hyperglycaemia. The second line of treatment is insulin therapy (GDMi). The latter protocol is reported to be effective in restoring glycaemia of the mother and the baby at birth. However, it is difficult to reach a consensus between the variety of protocols for insulin therapy since it depends on several factors including the population studied, ethnicity, among others. GDM*d* associates with deleterious effects on the foetoplacental vascular function, mainly due to endothelial dysfunction. These alterations regard with alterations in the L-arginine/nitric oxide signalling pathway, as well as in the expression of insulin receptors A and B, and insulin response. More recent studies suggest that c-Jun N-terminal kinase 1–mediated insulin resistance may result from increased endoplasmic reticulum stress in this type of cells from the human placenta. Interestingly, the insulin therapy is a protocol that does not restore the dysfunctional endothelium as seen in GDM*d*. Indeed, insulin therapy may associate with additional deleterious effects on the mother, the placenta and foetus, and the newborn in GDM. In this chapter, we summarised some examples of the wide variety of protocols for insulin therapy and the

* **Corresponding author Luis Sobrevia and Mario Subiabre:** Cellular and Molecular Physiology Laboratory (CMPL), Department of Obstetrics, Division of Obstetrics and Gynaecology, School of Medicine, Faculty of Medicine, Pontificia Universidad Católica de Chile, Santiago 8330024, Chile; E-mails: mesubiabre@uc.cl, lsobrevia@uc.cl

Atta-ur-Rahman (Ed.)

potential consequences of this protocol on the foetoplacental unit and the neonate from women with GDM.

Keywords: Diabetes, Diet, Endothelium, Endoplasmic reticulum stress, Gestational diabetes, Human, Insulin, Insulin therapy, Placenta.

INTRODUCTION

Gestational diabetes mellitus (GDM) is defined as any degree of glucose intolerance first recognised during pregnancy that is not overt diabetes [1]. The worldwide prevalence of this condition ranges from 6–20% of pregnant women [2, 3]. GDM shows with endothelial dysfunction and altered insulin signalling in the foetoplacental vasculature [4 - 10]. GDM is a disease of pregnancy that not only alters maternal metabolic parameters [11, 12] but also results in adverse foetal and newborn outcome. Worryingly, newborns to GDM pregnancies are prone to developing obesity and type 2 diabetes mellitus (T2DM) in adulthood [13 - 16].

Pregnant women diagnosed with GDM are first enrolled in a protocol including a controlled diet with regular glucose checking and physical activity (hereafter referred to as GDM*d*) [17]. The percentage of women with GDM that achieve suggested glycaemia values (*i.e.* fasting: $\leq$95 mg/dL (5.3 mmol/L), 1 h postprandial: $\leq$140 mg/dL (7.8 mmol/L), 2 h postprandial: $\leq$120 mg/dL (6.7 mmol/L)) [18] by changing their lifestyle is ~75% [1, 5, 14, 17, 19]. Women following a controlled diet but do not reach the recommended glycaemia values are referred to insulin therapy (*i.e.* GDM*i*) [1, 5, 20, 21]. Interestingly, insulin therapy in pregnant women with GDM seems to be equally efficient as diet [21]. It is reported that even when insulin therapy normalises maternal and newborn glycaemia the harmful effect of GDM on the placenta function and the neonate metabolic state persist [5, 22, 23]. Thus, other factors than plasma D-glucose may be involved in the effects of GDM on vascular function in the mother, the foetus, and the newborn [23]. One of the general proposed mechanisms that could explain this phenomenon include a metabolic memory as a phenomenon triggered by a short-term foetal exposure to high D-glucose or oscillating D-glucose level before and during GDM treatment [24].

Several metabolic alterations are seen in the foetoplacental vasculature in GDM, including abnormal metabolism of the endogenous nucleoside adenosine or the cationic amino acid L-arginine, and altered synthesis of nitric oxide (NO). These alterations are crucial in the regulation of the vascular function since these molecules are active vasodilators acting in concert in the placental vasculature in this disease of pregnancy [8 - 10]. Additionally, GDM associates with imbalanced unfolded protein response (UPR) leading to a state of endoplasmic reticulum (ER) stress. In ER stress, various factors are involved in the modulation of key phenomena that regulate the endothelial function. One of these altered signalling mechanisms is an increase in the expression of c-Jun N-terminal kinase 1 (JNK1), which associated with inhibition of insulin signalling [25]. Whether defective insulin actions in the foetoplacental vasculature in GDM are due to alterations in these mechanisms is unclear. In this review we summarised the current evidence about the treatment with insulin (*i.e.* insulin therapy) in pregnant women with GDM and the potential involvement of insulin signalling on the foetoplacental tissue.

Insulin Therapy in Gestational Diabetes Mellitus

The main goal of insulin therapy in GDM is to reduce the plasma glucose level to a normal range close to the glycaemia seen in pregnant women with a normal pregnancy. The final expected outcome is avoiding hyperglycaemia-associated maternal and foetal complications [5, 21, 26]. Several criteria are reported and used to decide the enrolment of pregnant women with GDM on a diet protocol or insulin therapy [5, 17, 21 - 23]. Individual studies suggest glycaemia values over which the insulin therapy should start. The values vary between 5.2–5.6 mmol/L (94–101 mg/dL) at fasting and between 6.6–7.9 mmol/L (119–142 mg/dL) after 2 h postprandial depending on the studied population [22, 27 - 29]. Efficiency of insulin therapy is expected to reach glycaemia values between 3.3–5 mmol/L (60–90 mg/dL) at fasting, 3.3–5.8 mmol/L (60–105 mg/dL) pre-prandial, 6.1–7.2 mmol/L (110–130 mg/dL) at 1 h post-prandial, 5.0–6.7 mmol/L (90–120 mg/dL) at 2 h post-prandial, and 3.3–6.7 mmol/L (60–120 mg/dL) at bedtime, with glycosylated haemoglobin A_{1c} (HbA_{1c}) within a normal range (≤6%) [30]. Unfortunately, one of the main conclusions recently reported for 7381 women with GDM proposed that there are not enough high-quality results to offer significant differences for health outcomes after using insulin in pregnant women with this condition [21, 23].

Protocols of Insulin Therapy

Insulin therapy in pregnant women with GDM refers to the use of neutral protamine Hagedorn (NPH) insulin. In general, a certain dosage of insulin should include 2 to 4 administrations daily. The rapid-acting insulin analogues lispro and aspart are continuously administered in patients that check their blood glucose level regularly and use glucose monitoring devices [31] including patients with type 1 diabetes mellitus (T1DM) [32]. Under this approach the insulin dosage is adjusted according to the variations of the glycaemia during the day. However, several different approaches showed that a proper decision for the administration of a given insulin dosage is highly dependent on the target population and local guidelines worldwide [17, 21, 33]. The time to start with the administrations of insulin during the day, in the evening, or at night are different (Fig. **1**). Also, the dosage varies along the different study groups [34] (Table **1**). For example, patients showing elevated morning fasting glucose should receive evening NPH insulin or those with elevated post-prandial glucose level may receive short-acting insulin at different dosage at breakfast, lunch, and dinner. Some perinatal guidelines suggest that insulin therapy should consider the characteristics of the woman and the frequency of self-monitored glycaemia [16, 35, 36].

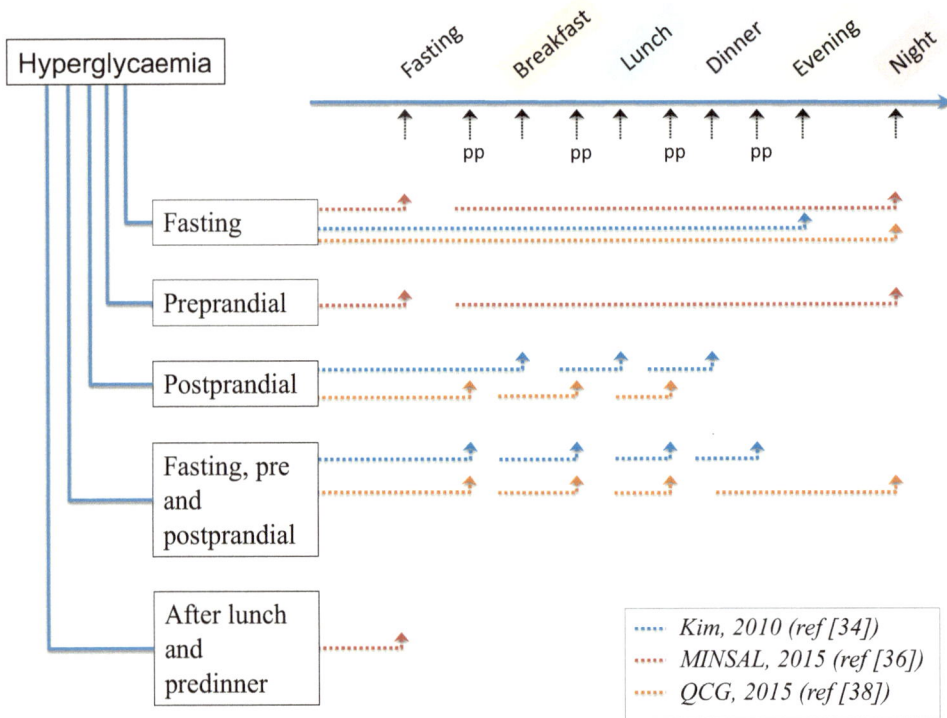

Fig. (1). Variability in the approach to initiate and develop insulin therapy in women with gestational diabetes mellitus. Insulin therapy is a protocol used in women with gestational diabetes mellitus subjected to a controlled diet but persistent hyperglycaemia. The several protocols for insulin therapy described in the literature show different starting points depending on the moment of the day when hyperglycaemia is detected. Thus, hyperglycaemia could be at fasting, pre-prandial or postprandial and insulin administration begins at fasting, pre-prandial or with the meals at breakfast, lunch, dinner, or even at bedtime at night. Dotted coloured arrows show some of the approaches suggested by the indicated references. pp is pre-prandial insulin administration.

Table 1. Insulin administration in women with gestational diabetes mellitus.

Altered Glycaemia	Insulin	Dosage	Time of Administration	Reference
Fasting	NPH	0.2 IU/kg	Evening	[34]
Post-prandial	Short-acting	1.5 IU per 10 g carbohydrate and 1 IU per 10 g carbohydrate	Breakfast and Lunch and dinner	[34]
Pre- and post-prandial	NPH	0.9-1.0 IU/ kg	Four injections per day	[34]
Fasting >100 mg glucose/dL	NPH	0.1 IU/kg/day	Night	[36]
Post-prandial breakfast 140-179 mg/dL	Rapid-acting	2 IU/kg	Pre-prandial (30 min before breakfast)	[36]
Post-prandial breakfast >180 mg/dL	Rapid-acting	4 IU/kg	Pre-prandial (30 min before breakfast)	[36]
Post-prandial lunch and before dinner	NPH	0.15 IU/kg	Before breakfast	[36]
Pre-prandial breakfast, lunch, and dinner	NPH	0.2 IU/kg	2/3 of dosage before breakfast and 1/3 at bedtime	[36]
ns	Regular and intermediate	1/3 and 2/3 of full dosage, respectively (*IU ns*)	Morning	[37]
ns	Regular	*ns*	Pre-prandial	[37]
ns	*ns*	0.1 IU/kg/day	*ns*	[82]
ns	Regular and NPH	0.7 IU/kg/day	2/3 dosage before breakfast and 1/3 dosage before dinner	[83]
ns	Regular and NPH	0.2 IU/kg/day	2/3 dosage of NPH in the morning and 1/3 dosage of Regular at evening	[84]
ns	Regular and NPH	0.8 IU/kg in the second and 0.9 IU/kg in the third trimester of pregnancy	Regular insulin before main meals and NPH at bed time	[85]
ns	NPH	0.8 UI/kg in the second and 0.9 UI/kg in the third trimester of pregnancy	*ns*	[86]

(Table 1) cont.....

Altered Glycaemia	Insulin	Dosage	Time of Administration	Reference
Fasting, pre-prandial or post-prandial	Lispro and regular	1 IU per 10 g carbohydrates	Regular insulin 15 min before and insulin lispro immediately before meals	[87]
ns	Regular and NPH	0.4 IU/kg (50% NPH and 50% Regular)	*ns*	[88]
ns	Regular and NPH	0.5 IU/kg	Twice daily (increased weekly if needed)	[89]

NPH, neutral protamine Hagedorn insulin; IU, international units; *ns*, not specified.

Patients with GDM are also subjected to two specific insulin treatment regimens [37], with the first approach consisting in a twice-daily insulin regimen, where the morning dose contained two thirds and the afternoon dose contained the remaining third of the total daily insulin. The morning dose consists in one third of human regular insulin (Actrapid, NovoNordisk, Denmark) and two thirds human intermediate insulin (Insulatard, NovoNordisk), and the evening dose consists in equal amounts of regular and intermediate insulin. According to the response that patients showed to insulin treatment adjustments are made in the insulin dosage according to each patient. The second approach is a four times daily insulin regimen where three doses of regular insulin are administered with insulin pen (Novopen 3, NovoNordisk) half an hour before each main meal, and the fourth dosage of intermediate insulin administered at bedtime.

Another example is the protocol implemented in The Queensland Clinical Guideline (QCG) [38] (Fig. **1**). The QCG suggests that the dose of insulin can be titrated every two or three days, depending on the response to treatment, increasing from 2 to 4 IU until reaching the glycaemic goals. Depending on the glycaemia different protocols are indicated. With an elevated fasting glucose, a single bedtime injection of intermediate-acting insulin is recommended. With postprandial hyperglycaemia, mealtime rapid acting insulin is required and whether there is fasting and postprandial hyperglycaemia a basal-bolus insulin regimen consisting in mealtime rapid-acting insulin and bedtime intermediate-acting or twice daily mixed insulin is suggested [38]. A recent alternative is the proposal from The American College of Obstetricians and Gynecologists (ACOG) [16]. ACOG recommends that to start insulin therapy it is necessary to calculate a total daily dosage of 0.7–1.0 IU per kg of body weight. Half of the total daily requirement is to administer as a single dose of long-acting insulin and the other half is given in three divided doses at mealtimes as rapid-acting insulin [16].

Based on the available literature and the above-mentioned protocols, it is evident that the use of insulin worldwide is far from a general approach with a specific insulin therapy plan. This is certainly dependent on factors including the ethnicity, sociocultural aspects such as type of food and drinks, physical activity, stress, and others [17, 21, 23, 33].

Insulin Therapy: Beneficial or Harmful to the Growing Foetus and Newborn?

Even when insulin therapy could put maternal and newborn glycaemia under control [21, 33] several alterations in the vascular metabolism persist at delivery altering the foetoplacental vascular function and increasing the risk of an adverse perinatal outcome (Fig. **2**). It is worth noting that women with GDM treated with diet or under insulin therapy show similar alterations in the foetoplacental endothelial function, suggesting that insulin therapy in GDM may not be better than diet or even than treating the women with oral anti-diabetic pharmacological therapies [17, 21, 23, 33].

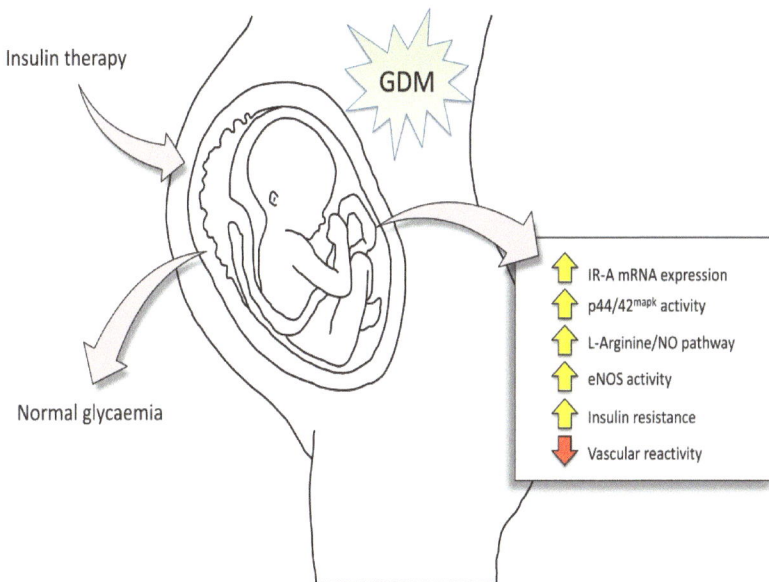

Fig. (2). Altered foetoplacental vascular function after insulin therapy in gestational diabetes mellitus. Pregnant women with diagnose of gestational diabetes mellitus (GDM) are subjected to treatment with insulin (Insulin therapy) resulting in normalization of the maternal and foetal glycaemia (Normal glycaemia). Several metabolic alterations in the foetoplacental vasculature are present at birth even after insulin therapy. These alterations include increased (↑) mRNA expression of the insulin receptor A (IR-A), 44 and 42 kDa mitogen-activated protein kinases (p44/42^{mapk}) activity, L-arginine transport and nitric oxide (NO) synthesis (L-Arginine/NO pathway), endothelial NO synthase (eNOS) activity, and insulin resistance; however, reduced (↓) the foetoplacental vascular reactivity.

The available information in studies for restricted population groups suggest that insulin therapy may be either beneficial or detrimental compared with women with GDM treated with diet, oral anti-diabetic drugs, or insulin analogues. Insulin therapy may reduce or not the incidence of macrosomia; however, subjecting pregnant women with GDM to insulin therapy associated with lower newborn birth weight when compared with women with GDM*d*, and lower newborn birth weight and higher ponderal index when compared with using an oral anti-diabetic pharmacological therapy [17, 21, 23, 33].

Restricted foetus growth results in foetoplacental vascular dysfunction at delivery [39] and a high ponderal index, suggestive of term growth status alterations [40], it is phenomena that associate with programming of adulthood diseases. Thus, insulin therapy to control maternal glycaemia results in deleterious effects for the growing foetus, newborn, infant, and perhaps their adult health status and quality of life. The scarcely reported studies comparing maternal insulin therapy *versus* diet in GDM show that diet could eventually be as good as insulin therapy or that insulin therapy does not improves foetal and newborn alterations caused by GDM compared with diet protocol [17, 21]. Thus, we should reconsider a protocol to reach a normal maternal glycaemia in pregnancy that will not have a consequence in the growing foetus and newborn [33, 41].

Other studies show that attention should be put to the use of long-lasting insulin in pregnant women with GDM since a higher incidence of macrosomia and newborn birthweight is reported when compared with the use of short-lasting insulin [42]. It is reported that the use of insulin will not always be beneficial compared with an oral anti-diabetic pharmacological therapy regarding newborn adverse outcomes [21, 33, 41]. For example, insulin therapy may result in supraphysiological maternal gestational weight gain and could alter infant weight at 12 to 18 months old causing a decrease in this parameter compared with women that were treated with metformin [43].

When comparing a hypoglycaemic oral drug such as glyburide *versus* insulin therapy, there were no differences between the groups about the percentage of newborns large for gestational age, with macrosomia, pulmonary complications, hypoglycaemia, or foetal abnormalities [44]. Equally, when comparing these two protocols there was no difference in the adiposity of newborns; however, the group treated with insulin presented lower macrosomia compared with the group treated with glyburide [45]. Other studies report the results of using metformin compared with insulin. Newborns to GDM pregnancies where the mother was under metformin showed lower circumference of the head, arm and thorax, and lower birth weight and height compared with those born to GDM where the mother was under insulin therapy [46]. Neonatal respiratory distress, neonatal jaundice, hyperbilirubinemia, admission to the neonatal intensive care unit, and premature birth was more frequent in the insulin therapy compared with metformin group [47]. In addition, the incidences of vacuum extraction deliveries and caesarean were significantly higher in the metformin group compared to the insulin therapy group [43]. Thus, diverse consequences of treating women with GDM with oral hypoglycaemic drugs or insulin therapy is reported [23], which makes uncertain the decision of which protocol follows in women with GDM.

Since supraphysiological maternal gestational weight gain is a key factor associated with dysfunction of the foetoplacental vasculature [48] it is worth-nothing that insulin therapy may cause foetoplacental vascular alterations [7, 23]. Additionally, the pre-pregnancy body mass index but not the normalization of glycaemia of the future mother is a potential predictor of newborn macrosomia in women with GDM under insulin therapy [49]. The described findings suggest that the variety of therapies applied in pregnant women with GDM are not enough to restore most of the parameters leading to an abnormal pregnancy outcome. Also, reaching maternal glycaemia in a normal range with different therapies in GDM is not enough to avoid the deleterious consequences in the neonate.

Mechanisms of Foetoplacental Dysfunction in GDM Under Insulin Therapy

The L-arginine/Nitric Oxide Signalling Pathway

L-Arginine is a cationic amino acid used by the endothelial NO synthase (eNOS) to synthesise L-citrulline generating NO as a coproduct in human umbilical vein endothelial cells (HUVECs) [50] and human placental microvascular endothelial cells (hPMECs) [51]. GDM associates with reduced human foetoplacental vascular response to insulin and adenosine involving decreased NOS activity. However, L-arginine transport mediated via the human cationic amino acid transporter 1 (CAT-1) isoform increased in HUVECs from GDM*d* [22, 52]. Interestingly, HUVECs from women with GDM*i* showed similar changes as in GDM*d*, suggesting that insulin therapy does not resolve the abnormal metabolic state of HUVECs in GDM (see Fig. **2**). The lack of a beneficial effect of maternal insulin therapy on the fetoplacental endothelium suggest that normalising the maternal and foetal glycaemia in response to insulin is not the unique or the most important factor involved in the GDM-associated alterations in the metabolism of the foetoplacental endothelium.

The level of NO is higher in HUVECs from GDM*d* and equally elevated in GDM*i* compared with cells from normal pregnancies [22]. Also, eNOS activation by higher activator phosphorylation of eNOS at serine[1177] is reported in HUVECs from GDM*d* and GDM*i*. Therefore, not only the level of NO but also eNOS activity is increased in the foetoplacental endothelium in this condition. Also, a potential scavenger of NO by overgeneration of free radicals, such as superoxide anion, is not enough or sufficient to mask the increased eNOS activity in HUVECs from GDM. However, the maximal capacity (defined as the inverse of the ratio between the half-maximal effect (EC_{50}) over maximal relaxation (R_{max}), $1/(EC_{50}/R_{max})$) of the human umbilical vein rings to respond to insulin and adenosine *in vitro* was significantly lower in GDM*i* ($1/(EC_{50}/R_{max})$ = 1.47) compared with GDM*d* ($1/(EC_{50}/R_{max})$ = 2.02) [22]. The latter is a phenomenon suggesting that insulin therapy worsened the response to insulin of these vessels compared with those from normal pregnancies. Furthermore, when insulin-induced dilation was estimated for endothelium-dependent response (*i.e.*, in response to calcitonin-gene related peptide as an endothelium preferential effector) [53], GDM*i* showed a more pronounced reduction in the maximal capacity of response of umbilical vein vessels compared with GDM*d* ($1/(EC_{50}/R_{max})$ = 8 *vs* 59 for GDM*i* *vs* GDM*d*, respectively). Therefore, insulin therapy may result in a preferential targeting to the endothelium in the foetoplacental vasculature in GDM. This response is another evidence suggesting that insulin therapy in GDM results in alterations that could even be more drastic than when women with GDM are treated only with diet.

Insulin Receptors and Signalling Pathways

At least two isoforms of insulin receptors (IRs) are reported [54]. These isoforms are generated through an alternative splicing process of the exon 11 in the insulin receptor gene *INSR*. The splicing generates a type of receptor lacking (IR-A) or having (IR-B) 12 amino acids at the C-terminal of the α-subunit [55 - 57]. Insulin activates IR-A and IR-B in target tissues in most cells [57], including foetal endothelium such as HUVECs [23, 50, 58] and hPMECs [51]. IR-A and IR-B activation by insulin results in the triggering of complex intracellular signalling pathways [8, 25]. IR-A activation causes a preferential activation of the signalling pathway mediated by 44 and 42 kDa mitogen-activated protein kinase (p44/42[mapk]) leading to a mitogenic phenotype [54, 59, 60]. IR-B activation preferentially activates the signalling pathways mediated by protein kinase B/Akt (Akt) leading to a metabolic phenotype [54, 59, 60]. Interestingly, adult endothelium as well as HUVECs and hPMECs respond to insulin with marked differences depending on the preferential expression of these subtypes of IRs [54]. Thus, insulin sensitivity and endothelium response to this hormone will depend on the expression as well as the signalling pathway activated by the IRs.

In GDM both the expression of IR-A and IR-B and their associated signalling pathways are altered in HUVECs and hPMECs [22, 50, 51, 58]. This phenomenon has been related to the endothelial dysfunction described in this pathology. HUVECs of GDM*d* showed higher number of mRNA copies determined for IR-A compared with cells from normal pregnancies, without significant alterations in the IR-B mRNA expression [22, 50, 58]. Consistent with these observations, the signalling pathways triggered by the activation of the IRs is also differential in this cell type. To date, GDMd associates with increased phosphorylation of p44/42^{mapk} (*i.e.* activation) compared with cells from normal cells. This phenomenon was not seen for Akt phosphorylation suggesting that altered IR-A rather than IR-B signalling in GDMd may be due to overexpression of IR-A [58]. In more recent studies this observation was extended to HUVECs from GDMi. The results described in these cells are similar to those seen in cells from GDMd [20]. Interestingly, the insulin treatment of pregnant women with GDM does not cause any correction in neither the GDM-associated increase of the IR-A mRNA expression nor the increase in p44/42^{mapk} activation. Thus, insulin therapy in women with GDM may not be effective in restoring the abnormal function of the fetoplacental endothelium at birth [23].

Endoplasmic Reticulum Stress

Abundant evidence establishes that women with GDM present insulin resistance in peripheral tissues [61]. However, the effect of GDM-associated insulin resistance in foetoplacental tissues and involved mechanisms are less documented [25, 62]. A reduced vasodilation caused by insulin in the umbilical vein from GDM [22] may be associated with a preferential decrease in the expression and activation of key proteins of the metabolic pathway (*i.e.* Akt signalling pathway) following IRs activation [15]. It is also described that ER stress plays a key role in the development of endothelial dysfunction and insulin resistance in metabolic syndrome-associated diseases [25, 63 - 65]. Therefore, ER stress could be a mechanism leading to deficient Akt-associated signalling in GDM [66].

ER stress results from the loss of the endoplasmic reticulum homeostasis and a subsequent accumulation of misfolding proteins ending in activation of the UPR. The UPR activation is aimed to restore the homeostasis of the endoplasmic reticulum and is characterised by the activation of three membrane proteins, *i.e.* protein kinase RNA-like endoplasmic reticulum kinase (PERK), inositol-requiring enzyme 1α (IRE1), and activating transcription factor 6 [25, 67]. One of the metabolic links between ER stress and insulin resistance is the IRE1-induced activation of JNK1. JNK1 directly phosphorylates insulin receptor substrate 1 at serine^{307} leading to its inactivation and a subsequent disruption of the IR-B/Ak--associated metabolic signalling branch of the insulin response [68]. In this regard, it is known that adipose tissue, skeletal muscle [69, 70], and placentas [71] from women with GDM presented ER stress, evidenced by higher activation of downstream proteins of UPR compared with women with normal pregnancies. In addition, the placentas from GDM pregnancies present with increased activation of JNK1. These findings suggest the occurrence of insulin resistance in placentas from GDM pregnancies. Moreover, HUVECs from GDM pregnancies show increased activity of C/EBP homologous protein (CHOP) [72], a PERK-downstream target protein, suggestive of ER stress in these endothelial cells. Since ER stress increases CAT-1 protein expression and L-arginine transport [65, 73, 74] and reduces the NO generation in HUVECs [75], ER stress could be involved in the lower insulin-induced vasodilation of umbilical veins and in the endothelial dysfunction described in foetoplacental endothelium from GDM [22]. The latter is also seen in HUVECs isolated from women with pre-pregnancy obesity, suggesting that obesity is a condition leading to ER stress [65]. Indeed, most studies showing alterations in the foetoplacental vasculature associated with GDM include women that are either overweight or obese. Thus, ER stress-associated vascular dysfunction could result from GDM, *i.e.* not T1DM or T2DM, detected in women with pre-pregnancy overweight or obesity, *i.e.* a condition recently referred to as 'gestational diabesity' [66, 76, 77]. GDM also associated with a systemic proinflammatory state [78]. In this regard, ER stress is a factor inducing the expression of proinflammatory cytokines such as tumour necrosis factor α (TNFα), and interleukins 6 and 1. TNFα causes activation of JNK1 thus contributing to inhibiting the metabolic insulin signalling pathway. Indeed, placenta tissue from GDM showed higher expression and release of TNFα [78, 79], suggesting a proinflammatory state that supports the presence of ER stress and a likely insulin resistance in this tissue. However, major studies addressing the involvement of inflammation in the foetoplacental tissue from GDM*d* and GDM*i* are required. Since insulin therapy reduced the expression of key proteins of the insulin signalling pathway in the human placenta compared with GDM*d* and normal pregnancies [15], suggests that insulin biological effects in the mother could result in modifying these alterations in foetus tissues. This phenomenon seems not to be the same in freshly isolated and cultured HUVECs from GDM*i* because not changes were detected in these cells compared with GDM*d* [22].

CONCLUDING REMARKS

The first line of treatment for GDM is diet control to minimize the impact of hyperglycaemia or to put the blood glucose values back to a normal physiological range [80]. In most cases this approach is effective, and women showed with normal glycaemia as well as their child at birth. However, a significant number of pregnant women under diet do not reach normal glycaemia and are passed to insulin therapy. As for diet treatment, insulin therapy is effective in normalising the maternal and newborn glycaemia. However, even when the plasma level of D-glucose is normalised after insulin therapy most of the described alterations in the foetoplacental vascular function in women treated with diet are still seen in pregnancies whether the mother was under insulin therapy. Several aspects are evident from these findings including the possibilities that (*i*) altered plasma D-glucose concentration may not be directly related with vascular dysfunction in the foetoplacental vasculature in GDM, (*ii*) since most of the women with GDM show pre-pregnancy overweight or obesity other factors than D-glucose may cause vascular alterations, including potential risk factors such as lipids (likely triglycerides) and adipocytes abnormal function, (*iii*) supraphysiological insulin resistance (pregnancy courses with a natural maternal and foetal insulin resistance state) in the foetoplacental vasculature in GDM, or (*iv*) other factors modulating the vascular endothelium metabolism including the L-arginine/NO signalling pathway or inducing ER stress may result in lower vascular response to insulin (Fig. **3**). These possibilities, among others, suggest potential mechanisms of insulin resistance that need to be unveiled in the foetoplacental vasculature.

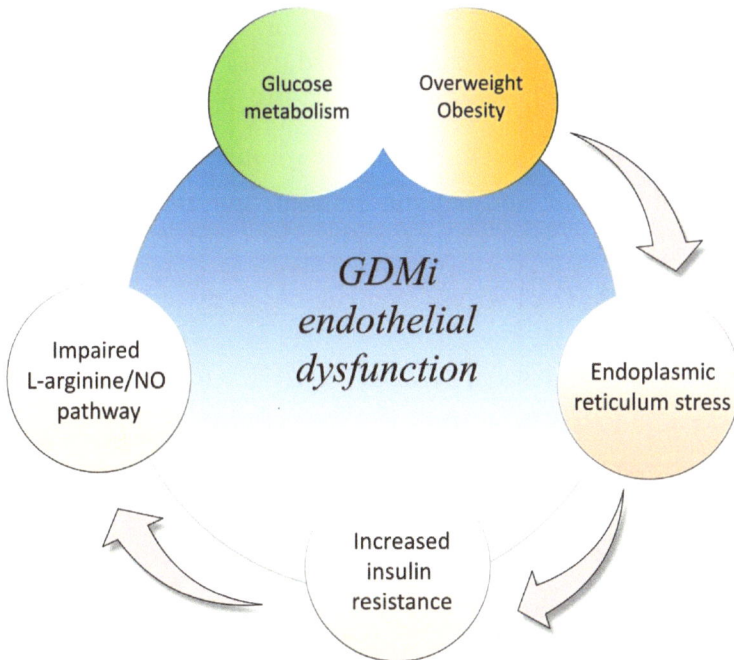

Fig. (3). Mechanisms of foetoplacental endothelial dysfunction in insulin-treated women with gestational diabetes mellitus. An altered metabolism of D-glucose (Glucose metabolism) is seen in women with pre-pregnancy obesity and overweight or women that develop these conditions during pregnancy. These pathological conditions associate with foetoplacental endothelial dysfunction in gestational diabetes mellitus where the mother was under insulin therapy (*GDMi endothelial dysfunction*) Overweight and obesity alters the metabolic state of the cells resulting in endoplasmic reticulum stress which makes the cells more susceptible to a defective insulin signalling in target cells causing an increase in the physiological insulin resistance seen in pregnancy (Increased insulin resistance). As a result of worsening the insulin resistance in GDM*i*, uptake of L-arginine and its metabolism to generate nitric oxide (NO) is impaired. These metabolic alterations and the corresponding specific mechanisms may act in concert to cause endothelial dysfunction in this disease of pregnancy.

Treatment with insulin seems not to be much more beneficial than a controlled diet in pregnant women with GDM, or even when compared with administration of anti-hyperglycaemic pharmacological drugs or changing the life style (physical activity, lower stress, *etc*) [17, 21, 23, 33, 81]. We face the fact that none of the protocols (diet, insulin therapy, drug therapy) seems to work to reduce the altered foetoplacental dysfunction seen in GDM. This is not, however, a negative outcome when diet is emphasized as an approach that results in a similar outcome than insulin therapy regarding GDM-associated alterations of the foetoplacental endothelial/vascular function. It is worth noting that diet may result in avoiding the several reported negative outcomes at the level of the mother, placenta and foetus, and newborn associated with insulin therapy in GDM. There is a considerable amount of evidence showing that several metabolic alterations are seen in the women, the foetus and placenta, and the newborn to GDM under insulin therapy are absent in women treated with diet, anti-hyperglycaemic pharmacological drugs, or with a change in the life style [17, 21 - 23, 33]. Thus, insulin therapy in GDM pregnancy is an approach to take with caution when the mother, the growing foetus and the newborn health may show alterations as a result of the insulin treatment beyond normalising the maternal and foetal glycaemia.

CONSENT FOR PUBLICATION

Not applicable.

CONFLICT OF INTEREST

The author confirms that this chapter contents have no conflict of interest.

ACKNOWLEDGEMENTS

Support was from the Fondo Nacional de Desarrollo Científico y Tecnológico (FONDECYT 1190316, 1150377, 11150083), Chile. This project has received funding from the Marie Curie International Research Staff Exchange Scheme with the 7th European Community Framework Program [grant agreement number 295185 - EULAMDIMA]. MS, RV-L, and LSi hold Comisión Nacional para la Investigación en Ciencia y Tecnología (CONICYT) and Vicerrectorate of Research, PUC (Chile)-PhD fellowships. LSi holds PhD fellowships from The Abel Tasman Talent Program and University Medical Center Groningen (UMCG) (The Netherlands) and School of Medicine, PUC (Chile).

REFERENCES

[1]   American Diabetes Association. Management of Diabetes in Pregnancy: *Standards of Medical Care in Diabetes-2018.* Diabetes Care 2018; 41 (Suppl. 1): S137-43.
[http://dx.doi.org/10.2337/dc18-S013] [PMID: 29222384]

[2]   Melchior H, Kurch-Bek D, Mund M. The prevalence of gestational diabetes. Dtsch Arztebl Int 2017; 114(24): 412-8.
       [PMID: 28669379]

[3]   Egan AM, Vellinga A, Harreiter J, *et al.* Epidemiology of gestational diabetes mellitus according to IADPSG/WHO 2013 criteria among obese pregnant women in Europe. Diabetologia 2017; 60(10): 1913-21.
       [http://dx.doi.org/10.1007/s00125-017-4353-9] [PMID: 28702810]

[4]   Haas TL. Shaping and remodeling of the fetoplacental circulation: aspects of health and disease. Microcirculation 2014; 21(1): 1-3.
       [http://dx.doi.org/10.1111/micc.12084] [PMID: 24033789]

[5]   Sobrevia L, Salsoso R, Sáez T, Sanhueza C, Pardo F, Leiva A. Insulin therapy and fetoplacental vascular function in gestational diabetes mellitus. Exp Physiol 2015; 100(3): 231-8.
       [http://dx.doi.org/10.1113/expphysiol.2014.082743] [PMID: 25581778]

[6]   Wagner R, Fritsche L, Heni M, *et al.* A novel insulin sensitivity index particularly suitable to measure insulin sensitivity during gestation. Acta Diabetol 2016; 53(6): 1037-44.
       [http://dx.doi.org/10.1007/s00592-016-0930-5] [PMID: 27771766]

[7]   Lapolla A, Dalfrà MG, Ragazzi E, De Cata AP, Fedele D. New International Association of the Diabetes and Pregnancy Study Groups (IADPSG) recommendations for diagnosing gestational diabetes compared with former criteria: a retrospective study on pregnancy outcome. Diabet Med 2011; 28(9): 1074-7.
       [http://dx.doi.org/10.1111/j.1464-5491.2011.03351.x] [PMID: 21658125]

[8]   Silva L, Subiabre M, Araos J, *et al.* Insulin/adenosine axis linked signalling. Mol Aspects Med 2017; 55: 45-61.
       [http://dx.doi.org/10.1016/j.mam.2016.11.002] [PMID: 27871900]

[9]   Sáez T, de Vos P, Sobrevia L, Faas MM. Is there a role for exosomes in foetoplacental endothelial dysfunction in gestational diabetes mellitus? Placenta 2018; 61: 48-54.
       [http://dx.doi.org/10.1016/j.placenta.2017.11.007] [PMID: 29277271]

[10]   Sáez T, Salsoso R, Leiva A, *et al.* Human umbilical vein endothelium-derived exosomes play a role in foetoplacental endothelial dysfunction in gestational diabetes mellitus. Biochim Biophys Acta Mol Basis Dis 2018; 1864(2): 499-508.
       [http://dx.doi.org/10.1016/j.bbadis.2017.11.010] [PMID: 29155213]

[11]   Law KP, Zhang H. The pathogenesis and pathophysiology of gestational diabetes mellitus: Deductions from a three-part longitudinal metabolomics study in China. Clin Chim Acta 2017; 468: 60-70.
       [http://dx.doi.org/10.1016/j.cca.2017.02.008] [PMID: 28213010]

[12]   Law KP, Mao X, Han TL, Zhang H. Unsaturated plasma phospholipids are consistently lower in the patients diagnosed with gestational diabetes mellitus throughout pregnancy: A longitudinal metabolomics study of Chinese pregnant women part 1. Clin Chim Acta 2017; 465: 53-71.
       [http://dx.doi.org/10.1016/j.cca.2016.12.010] [PMID: 27988319]

[13]   Catalano PM, Kirwan JP, Haugel-de Mouzon S, King J. Gestational diabetes and insulin resistance: role in short- and long-term implications for mother and fetus. J Nutr 2003; 133(5) (Suppl. 2): 1674S-83S.
       [http://dx.doi.org/10.1093/jn/133.5.1674S] [PMID: 12730484]

[14]   Metzger BE, Buchanan TA, Coustan DR, *et al.* Summary and recommendations of the fifth international workshop-conference on gestational diabetes mellitus. Diabetes Care 2007; 30 (Suppl. 2): S251-60.
       [http://dx.doi.org/10.2337/dc07-s225] [PMID: 17596481]

[15]   Colomiere M, Permezel M, Riley C, Desoye G, Lappas M. Defective insulin signaling in placenta from pregnancies complicated by gestational diabetes mellitus. Eur J Endocrinol 2009; 160(4): 567-78.

[http://dx.doi.org/10.1530/EJE-09-0031] [PMID: 19179458]

[16]   Practice Bulletin No. 180. Obstet Gynecol 2017; 130: 17-37.
       [http://dx.doi.org/10.1097/AOG.0000000000002159]

[17]   Brown J, Alwan NA, West J, *et al.* Lifestyle interventions for the treatment of women with gestational
       diabetes. Cochrane Database Syst Rev 2017; 5CD011970 a
       [http://dx.doi.org/10.1002/14651858.CD011970.pub2] [PMID: 28472859]

[18]   Classification and diagnosis of diabetes. Diabetes Care 2016; 39: 13-22.
       [http://dx.doi.org/10.2337/dc16-S005]

[19]   Mayo K, Melamed N, Vandenberghe H, Berger H. The impact of adoption of the international
       association of diabetes in pregnancy study group criteria for the screening and diagnosis of gestational
       diabetes. Am J Obstet Gynecol 2015; 212(2): 224.e1-9.
       [http://dx.doi.org/10.1016/j.ajog.2014.08.027] [PMID: 25173183]

[20]   Hartling L, Dryden DM, Guthrie A, Muise M, Vandermeer B, Donovan L. Benefits and harms of
       treating gestational diabetes mellitus: a systematic review and meta-analysis for the U.S. Preventive
       Services Task Force and the National Institutes of Health Office of Medical Applications of Research.
       Ann Intern Med 2013; 159(2): 123-9.
       [http://dx.doi.org/10.7326/0003-4819-159-2-201307160-00661] [PMID: 23712381]

[21]   Brown J, Grzeskowiak L, Williamson K, Downie MR, Crowther CA. Insulin for the treatment of
       women with gestational diabetes. Cochrane Database Syst Rev 2017; 11CD012037 b
       [http://dx.doi.org/10.1002/14651858.CD012037.pub2] [PMID: 29103210]

[22]   Subiabre M, Silva L, Villalobos-Labra R, *et al.* Maternal insulin therapy does not restore
       foetoplacental endothelial dysfunction in gestational diabetes mellitus. Biochim Biophys Acta Mol
       Basis Dis 2017; 1863(11): 2987-98.
       [http://dx.doi.org/10.1016/j.bbadis.2017.07.022] [PMID: 28756217]

[23]   Subiabre M, Silva L, Toledo F, *et al.* Insulin therapy and its consequences for the mother, foetus, and
       newborn in gestational diabetes mellitus. Biochim Biophys Acta Mol Basis Dis 2018; 1864(9 Pt B):
       2949-56.
       [http://dx.doi.org/10.1016/j.bbadis.2018.06.005] [PMID: 29890222]

[24]   Schisano B, Tripathi G, McGee K, McTernan PG, Ceriello A. Glucose oscillations, more than constant
       high glucose, induce p53 activation and a metabolic memory in human endothelial cells. Diabetologia
       2011; 54(5): 1219-26.
       [http://dx.doi.org/10.1007/s00125-011-2049-0] [PMID: 21287141]

[25]   Yung HW, Alnæs-Katjavivi P, Jones CJP, *et al.* Placental endoplasmic reticulum stress in gestational
       diabetes: the potential for therapeutic intervention with chemical chaperones and antioxidants.
       Diabetologia 2016; 59(10): 2240-50.
       [http://dx.doi.org/10.1007/s00125-016-4040-2] [PMID: 27406815]

[26]   Villalobos-Labra R, Silva L, Subiabre M, *et al.* Akt/mTOR Role in human foetoplacental vascular
       insulin resistance in diseases of pregnancy. J Diabetes Res 2017; 20175947859
       [http://dx.doi.org/10.1155/2017/5947859] [PMID: 29104874]

[27]   Jacqueminet S, Jannot-Lamotte MF. Therapeutic management of gestational diabetes. Diabetes Metab
       2010; 36(6 Pt 2): 658-71.
       [http://dx.doi.org/10.1016/j.diabet.2010.11.016] [PMID: 21163428]

[28]   Botta RM, Di Giovanni BM, Cammilleri F, Taravella V. Predictive factors for insulin treatment in
       women with diagnosis of gestational diabetes. Ann Ist Super Sanita 1997; 33(3): 403-6.
       [PMID: 9542271]

[29]   Crowther CA, Hiller JE, Moss JR, McPhee AJ, Jeffries WS, Robinson JS. Effect of treatment of
       gestational diabetes mellitus on pregnancy outcomes. N Engl J Med 2005; 352(24): 2477-86.
       [http://dx.doi.org/10.1056/NEJMoa042973] [PMID: 15951574]

[30]   Landon MB, Spong CY, Thom E, *et al.* A multicenter, randomized trial of treatment for mild gestational diabetes. N Engl J Med 2009; 361(14): 1339-48.
[http://dx.doi.org/10.1056/NEJMoa0902430] [PMID: 19797280]

[31]   Lapolla A, Dalfrà MG, Fedele D. Insulin therapy in pregnancy complicated by diabetes: are insulin analogs a new tool? Diabetes Metab Res Rev 2005; 21(3): 241-52.
[http://dx.doi.org/10.1002/dmrr.551] [PMID: 15818714]

[32]   Hirsch IB. Insulin analogues. N Engl J Med 2005; 352(2): 174-83.
[http://dx.doi.org/10.1056/NEJMra040832] [PMID: 15647580]

[33]   Bhattacharyya A, Brown S, Hughes S, Vice PA. Insulin lispro and regular insulin in pregnancy. QJM 2001; 94(5): 255-60.
[http://dx.doi.org/10.1093/qjmed/94.5.255] [PMID: 11353099]

[34]   Kim C. Gestational diabetes: risks, management, and treatment options. Int J Womens Health 2010; 2: 339-51.
[http://dx.doi.org/10.2147/IJWH.S13333] [PMID: 21151681]

[35]   Committee on Practice Bulletins--Obstetrics. Practice Bulletin No. 137: Gestational diabetes mellitus. Obstet Gynecol 2013; 122(2 Pt 1): 406-16.
[PMID: 23969827]

[36]   Guía Perinatal. 2015.web.minsal.cl/wp-content/uploads/2015/10/GUIA-PERINATAL_2015.10.08_ web.pdf-R.pdf

[37]   Nachum Z, Ben-Shlomo I, Weiner E, Shalev E. Twice daily *versus* four times daily insulin dose regimens for diabetes in pregnancy: randomised controlled trial. BMJ 1999; 319(7219): 1223-7.
[http://dx.doi.org/10.1136/bmj.319.7219.1223] [PMID: 10550081]

[38]   Gestation diabetes mellitus. Queensl Matern Neonatal Clin Guidel 2015; 2015: 1-38.

[39]   Casanello P, Escudero C, Sobrevia L. Equilibrative nucleoside (ENTs) and cationic amino acid (CATs) transporters: implications in foetal endothelial dysfunction in human pregnancy diseases. Curr Vasc Pharmacol 2007; 5(1): 69-84.
[http://dx.doi.org/10.2174/157016107779317198] [PMID: 17266615]

[40]   Khoury MJ, Berg CJ, Calle EE. The ponderal index in term newborn siblings. Am J Epidemiol 1990; 132(3): 576-83.
[http://dx.doi.org/10.1093/oxfordjournals.aje.a115694] [PMID: 2389761]

[41]   Bogdanet D, Egan A, Reddin C, Kirwan B, Carmody L, Dunne F. ATLANTIC DIP: Despite insulin therapy in women with IADPSG diagnosed GDM, desired pregnancy outcomes are still not achieved. What are we missing? Diabetes Res Clin Pract 2018; 136: 116-23.
[http://dx.doi.org/10.1016/j.diabres.2017.12.003] [PMID: 29253626]

[42]   Pöyhönen-Alho M, Teramo K, Kaaja R. Treatment of gestational diabetes with short- or long-acting insulin and neonatal outcome: a pilot study. Acta Obstet Gynecol Scand 2002; 81(3): 258-9.
[http://dx.doi.org/10.1034/j.1600-0412.2002.810312.x] [PMID: 11966484]

[43]   Ijäs H, Vääräsmäki M, Morin-Papunen L, *et al.* Metformin should be considered in the treatment of gestational diabetes: a prospective randomised study. BJOG 2011; 118(7): 880-5.
[http://dx.doi.org/10.1111/j.1471-0528.2010.02763.x] [PMID: 21083860]

[44]   Langer O, Conway DL, Berkus MD, Xenakis EM-J, Gonzales O. A comparison of glyburide and insulin in women with gestational diabetes mellitus. N Engl J Med 2000; 343(16): 1134-8.
[http://dx.doi.org/10.1056/NEJM200010193431601] [PMID: 11036118]

[45]   Lain KY, Garabedian MJ, Daftary A, Jeyabalan A. Neonatal adiposity following maternal treatment of gestational diabetes with glyburide compared with insulin. Am J Obstet Gynecol 2009; 200(5): 501.e1-6.
[http://dx.doi.org/10.1016/j.ajog.2009.02.038] [PMID: 19375570]

[46]   Niromanesh S, Alavi A, Sharbaf FR, Amjadi N, Moosavi S, Akbari S. Metformin compared with insulin in the management of gestational diabetes mellitus: a randomized clinical trial. Diabetes Res Clin Pract 2012; 98(3): 422-9.
[http://dx.doi.org/10.1016/j.diabres.2012.09.031] [PMID: 23068960]

[47]   Mesdaghinia E, Samimi M, Homaei Z, Saberi F, Moosavi SGA, Yaribakht M. Comparison of newborn outcomes in women with gestational diabetes mellitus treated with metformin or insulin: a randomised blinded trial. Int J Prev Med 2013; 4(3): 327-33.
[PMID: 23626890]

[48]   Pardo F, Silva L, Sáez T, *et al.* Human supraphysiological gestational weight gain and fetoplacental vascular dysfunction. Int J Obes 2015; 39(8): 1264-73.
[http://dx.doi.org/10.1038/ijo.2015.57] [PMID: 25869606]

[49]   Olmos PR, Borzone GR, Olmos RI, *et al.* Gestational diabetes and pre-pregnancy overweight: possible factors involved in newborn macrosomia. J Obstet Gynaecol Res 2012; 38(1): 208-14.
[http://dx.doi.org/10.1111/j.1447-0756.2011.01681.x] [PMID: 22070342]

[50]   Westermeier F, Salomón C, González M, *et al.* Insulin restores gestational diabetes mellitus-reduced adenosine transport involving differential expression of insulin receptor isoforms in human umbilical vein endothelium. Diabetes 2011; 60(6): 1677-87.
[http://dx.doi.org/10.2337/db11-0155] [PMID: 21515851]

[51]   Salomón C, Westermeier F, Puebla C, *et al.* Gestational diabetes reduces adenosine transport in human placental microvascular endothelium, an effect reversed by insulin. PLoS One 2012; 7(7): e40578.
[http://dx.doi.org/10.1371/journal.pone.0040578] [PMID: 22808198]

[52]   Guzmán-Gutiérrez E, Armella A, Toledo F, Pardo F, Leiva A, Sobrevia L. Insulin requires A1 adenosine receptors expression to reverse gestational diabetes-increased L-arginine transport in human umbilical vein endothelium. Purinergic Signal 2016; 12(1): 175-90.
[http://dx.doi.org/10.1007/s11302-015-9491-2] [PMID: 26710791]

[53]   Wang X, Han C, Fiscus RR. Calcitonin gene-related peptide (CGRP) causes endothelium-dependent cyclic AMP, cyclic GMP and vasorelaxant responses in rat abdominal aorta. Neuropeptides 1991; 20(2): 115-24.
[http://dx.doi.org/10.1016/0143-4179(91)90061-M] [PMID: 1724683]

[54]   Westermeier F, Sáez T, Arroyo P, *et al.* Insulin receptor isoforms: an integrated view focused on gestational diabetes mellitus. Diabetes Metab Res Rev 2016; 32(4): 350-65.
[http://dx.doi.org/10.1002/dmrr.2729] [PMID: 26431063]

[55]   Ullrich A, Bell JR, Chen EY, *et al.* Human insulin receptor and its relationship to the tyrosine kinase family of oncogenes. Nature 1985; 313(6005): 756-61.
[http://dx.doi.org/10.1038/313756a0] [PMID: 2983222]

[56]   Ebina Y, Edery M, Ellis L, *et al.* Expression of a functional human insulin receptor from a cloned cDNA in Chinese hamster ovary cells. Proc Natl Acad Sci USA 1985; 82(23): 8014-8.
[http://dx.doi.org/10.1073/pnas.82.23.8014] [PMID: 3906655]

[57]   Belfiore A, Frasca F, Pandini G, Sciacca L, Vigneri R. Insulin receptor isoforms and insulin receptor/insulin-like growth factor receptor hybrids in physiology and disease. Endocr Rev 2009; 30(6): 586-623.
[http://dx.doi.org/10.1210/er.2008-0047] [PMID: 19752219]

[58]   Westermeier F, Salomón C, Farías M, *et al.* Insulin requires normal expression and signaling of insulin receptor A to reverse gestational diabetes-reduced adenosine transport in human umbilical vein endothelium. FASEB J 2015; 29(1): 37-49.
[http://dx.doi.org/10.1096/fj.14-254219] [PMID: 25351985]

[59]   Arroyo P, Guzmán-Gutiérrez E, Pardo F, *et al.* Gestational diabetes mellitus and the role of adenosine in the human placental endothelium and central nervous system. Glob J Pathol Microbiol 2013; 1: 24-

42.

[60]   Muniyappa R, Sowers JR. Role of insulin resistance in endothelial dysfunction. Rev Endocr Metab Disord 2013; 14(1): 5-12.
[http://dx.doi.org/10.1007/s11154-012-9229-1] [PMID: 23306778]

[61]   Barbour LA, McCurdy CE, Hernandez TL, Kirwan JP, Catalano PM, Friedman JE. Cellular mechanisms for insulin resistance in normal pregnancy and gestational diabetes. Diabetes Care 2007; 30 (Suppl. 2): S112-9.
[http://dx.doi.org/10.2337/dc07-s202] [PMID: 17596458]

[62]   Ruiz-Palacios M, Ruiz-Alcaraz AJ, Sanchez-Campillo M, Larqué E. Role of insulin in placental transport of nutrients in gestational diabetes mellitus. Ann Nutr Metab 2017; 70(1): 16-25.
[http://dx.doi.org/10.1159/000455904] [PMID: 28110332]

[63]   Battson ML, Lee DM, Gentile CL. Endoplasmic reticulum stress and the development of endothelial dysfunction. Am J Physiol Heart Circ Physiol 2017; 312(3): H355-67.
[http://dx.doi.org/10.1152/ajpheart.00437.2016] [PMID: 27923788]

[64]   Cnop M, Foufelle F, Velloso LA. Endoplasmic reticulum stress, obesity and diabetes. Trends Mol Med 2012; 18(1): 59-68.
[http://dx.doi.org/10.1016/j.molmed.2011.07.010] [PMID: 21889406]

[65]   Villalobos-Labra R, Sáez PJ, Subiabre M, *et al.* Pre-pregnancy maternal obesity associates with endoplasmic reticulum stress in human umbilical vein endothelium. Biochim Biophys Acta Mol Basis Dis 2018; 1864(10): 3195-210.
[http://dx.doi.org/10.1016/j.bbadis.2018.07.007] [PMID: 30006153]

[66]   Villalobos-Labra R, Subiabre M, Toledo F, Pardo F, Sobrevia L. Endoplasmic reticulum stress and development of insulin resistance in adipose, skeletal, liver, and foetoplacental tissue in diabesity. Mol Aspects Med 2019; 66: 49-61.
[http://dx.doi.org/10.1016/j.mam.2018.11.001] [PMID: 30472165]

[67]   Flamment M, Hajduch E, Ferré P, Foufelle F. New insights into ER stress-induced insulin resistance. Trends Endocrinol Metab 2012; 23(8): 381-90.
[http://dx.doi.org/10.1016/j.tem.2012.06.003] [PMID: 22770719]

[68]   Ozcan U, Cao Q, Yilmaz E, *et al.* Endoplasmic reticulum stress links obesity, insulin action, and type 2 diabetes. Science 2004; 306(5695): 457-61.
[http://dx.doi.org/10.1126/science.1103160] [PMID: 15486293]

[69]   Liong S, Lappas M. Endoplasmic reticulum stress is increased in adipose tissue of women with gestational diabetes. PLoS One 2015; 10(4)e0122633
[http://dx.doi.org/10.1371/journal.pone.0122633] [PMID: 25849717]

[70]   Liong S, Lappas M. Endoplasmic reticulum stress regulates inflammation and insulin resistance in skeletal muscle from pregnant women. Mol Cell Endocrinol 2016; 425: 11-25.
[http://dx.doi.org/10.1016/j.mce.2016.02.016] [PMID: 26902174]

[71]   Farías M, Puebla C, Westermeier F, *et al.* Nitric oxide reduces SLC29A1 promoter activity and adenosine transport involving transcription factor complex hCHOP-C/EBPalpha in human umbilical vein endothelial cells from gestational diabetes. Cardiovasc Res 2010; 86(1): 45-54.
[http://dx.doi.org/10.1093/cvr/cvp410] [PMID: 20032083]

[72]   Huang CC, Li Y, Lopez AB, *et al.* Temporal regulation of Cat-1 (cationic amino acid transporter-1) gene transcription during endoplasmic reticulum stress. Biochem J 2010; 429(1): 215-24.
[http://dx.doi.org/10.1042/BJ20100286] [PMID: 20408811]

[73]   Lopez AB, Wang C, Huang CC, *et al.* A feedback transcriptional mechanism controls the level of the arginine/lysine transporter cat-1 during amino acid starvation. Biochem J 2007; 402(1): 163-73.
[http://dx.doi.org/10.1042/BJ20060941] [PMID: 17042743]

[74]   Murugan D, Lau YS, Lau CW, Mustafa MR, Huang Y. Angiotensin 1-7 protects against angiotensin

II-induced endoplasmic reticulum stress and endothelial dysfunction *via* mas receptor. PLoS One 2015; 10(12)e0145413
[http://dx.doi.org/10.1371/journal.pone.0145413] [PMID: 26709511]

[75]   Desoye G, Hauguel-de Mouzon S. The human placenta in gestational diabetes mellitus. The insulin and cytokine network. Diabetes Care 2007; 30 (Suppl. 2): S120-6.
[http://dx.doi.org/10.2337/dc07-s203] [PMID: 17596459]

[76]   Cabalín C, Villalobos-Labra R, Toledo F, Sobrevia L. Involvement of A2B adenosine receptors as anti-inflammatory in gestational diabesity. Mol Aspects Med 2019; 66: 31-9.
[http://dx.doi.org/10.1016/j.mam.2019.01.001] [PMID: 30664911]

[77]   Pardo F, Subiabre M, Fuentes G, *et al.* Altered foetoplacental vascular endothelial signalling to insulin in diabesity. Mol Aspects Med 2019; 66: 40-8.
[http://dx.doi.org/10.1016/j.mam.2019.02.003] [PMID: 30849412]

[78]   Coughlan MT, Oliva K, Georgiou HM, Permezel JMH, Rice GE. Glucose-induced release of tumour necrosis factor-alpha from human placental and adipose tissues in gestational diabetes mellitus. Diabet Med 2001; 18(11): 921-7.
[http://dx.doi.org/10.1046/j.1464-5491.2001.00614.x] [PMID: 11703438]

[79]   World Health Organization. Guidelines on second-and third-line medicines and type of insulin for the control of blood glucose levels in non-pregnant adults with diabetes mellitus 2018.

[80]   Vérier-Mine O. Outcomes in women with a history of gestational diabetes. Screening and prevention of type 2 diabetes. Literature review. Diabetes Metab 2010; 36(6 Pt 2): 595-616.
[http://dx.doi.org/10.1016/j.diabet.2010.11.011] [PMID: 21163424]

[81]   Tieu J, Shepherd E, Middleton P, Crowther CA. Dietary advice interventions in pregnancy for preventing gestational diabetes mellitus. Cochrane Database Syst Rev 2017; 1CD006674
[http://dx.doi.org/10.1002/14651858.CD006674.pub3] [PMID: 28046205]

[82]   Anjalakshi C, Balaji V, Balaji MS, Seshiah V. A prospective study comparing insulin and glibenclamide in gestational diabetes mellitus in Asian Indian women. Diabetes Res Clin Pract 2007; 76(3): 474-5.
[http://dx.doi.org/10.1016/j.diabres.2006.09.031] [PMID: 17113179]

[83]   Ashoush S, El-Said M, Fathi H, Abdelnaby M. Identification of metformin poor responders, requiring supplemental insulin, during randomization of metformin *versus* insulin for the control of gestational diabetes mellitus. J Obstet Gynaecol Res 2016; 42(6): 640-7.
[http://dx.doi.org/10.1111/jog.12950] [PMID: 26992090]

[84]   Behrashi M, Samimi M, Ghasemi T, Saberi F, Atoof F. Comparison of glibenclamide and insulin on neonatal outcomes in pregnant women with gestational diabetes. Int J Prev Med 2016; 7: 88.
[http://dx.doi.org/10.4103/2008-7802.184502] [PMID: 27413519]

[85]   Bertini AM, Silva JC, Taborda W, *et al.* Perinatal outcomes and the use of oral hypoglycemic agents. J Perinat Med 2005; 33(6): 519-23.
[http://dx.doi.org/10.1515/JPM.2005.092] [PMID: 16318615]

[86]   De Veciana M, Trail PA, Evans AT, Dulaney K. A comparison of oral acarbose and insulin in women with gestational diabetes. Obstet Gynecol 2002; 99(4) (Suppl.): 5S.
[http://dx.doi.org/10.1016/S0029-7844(02)01676-9]

[87]   Mecacci F, Carignani L, Cioni R, *et al.* Maternal metabolic control and perinatal outcome in women with gestational diabetes treated with regular or lispro insulin: comparison with non-diabetic pregnant women. Eur J Obstet Gynecol Reprod Biol 2003; 111(1): 19-24.
[http://dx.doi.org/10.1016/S0301-2115(03)00157-X] [PMID: 14557006]

[88]   Mirzamoradi M, Heidar Z, Faalpoor Z, Naeiji Z, Jamali R. Comparison of glyburide and insulin in women with gestational diabetes mellitus and associated perinatal outcome: a randomized clinical trial. Acta Med Iran 2015; 53(2): 97-103.

[PMID: 25725178]

[89]   Zangeneh M, Veisi F, Ebrahimi B, Rezavand N. Comparison of therapeutic effects of insulin and glibenclamide in gestational diabetes. IJOGI 2014; 17: 1-7.

CHAPTER 3

Insights on Diabetes, Oxidative Stress and Antioxidant Therapeutic Strategies

Nicolette Nadene Houreld[*] and **Naresh Kumar Rajendran**

Laser Research Centre, Faculty of Health Sciences, University of Johannesburg, Johannesburg, 2028, South Africa

Abstract: Diabetes mellitus (DM) is a serious health concern that affects millions of people worldwide. Despite numerous studies on the topic, the exact mechanisms underlying diabetes progression and its complications is still unclear. Growing evidence suggests that hyperglycemia results in increased reactive oxygen species (ROS) production, leading to oxidative stress which affects and damages various tissues and organs. Oxidative stress results from an imbalance between ROS and antioxidants. During cellular metabolism free radicals such as ROS and reactive nitrogen species (RNS) are produced, and these free radicals have dual effects (both positive and negative) on nearby tissues and activate several oxidative stress-related signaling pathways. Oxidative stress has been identified as a major player in the pathogenesis of diabetes and its associated complications such as stroke, neuropathy, retinopathy, peripheral vascular disease, nephropathy and lower limb ulceration. Oxidative stress damages the surrounding tissue, and the effects continue for extended periods even after blood glucose concentrations return to normal. Prolonged oxidative stress results in insulin resistance, β-cell dysfunction, glucose intolerance and mitochondrial damage. Antioxidants are a group of enzymatic or non-enzymatic molecules that encounter and neutralize free radicals, thereby protecting the body from oxidative stress. Many exogenous molecules such as antioxidant supplements, vitamins (vitamin C and E) and metal ion chelators detoxify free radicals and maintain physiological levels. A better understanding of the involvement of oxidative stress in the pathogenesis of diabetes could have major therapeutic implications for treatment. An effective approach to treat oxidative stress is by using exogenous drugs that mimic antioxidants. Overall, this chapter highlights the understanding of oxidative stress-related mechanisms underlying the development of diabetes. It also elaborates on antioxidant therapy strategies to diminish oxidative stress and to treat diabetic associated complications.

Keywords: Antioxidants, Catalase, Diabetes, Free radicals, Glutathione, Hyperglycemia, Oxidative stress, Reactive oxygen species, Reactive nitrogen species, Superoxide dismutase.

---

[*] **Corresponding author Nicolette Nadene Houreld:** Laser Research Centre, Faculty of Health Sciences, University of Johannesburg, Johannesburg, 2028, South Africa; Tel: (+27) 11 559 6833; Fax: (+27) 11 559 6884; E-mail: nhoureld@uj.ac.za

Atta-ur-Rahman (Ed.)

1. INTRODUCTION

Diabetes mellitus (DM) is a lifelong chronic, metabolic, non-communicable disease (NCD) characterized by prolonged hyperglycemia due to impaired insulin secretion/utilization and insulin metabolism, and is associated with defects in the metabolism of carbohydrates, lipids and proteins. Complications associated with DM vary from person to person and are determined by an individual's health and lifestyle [1, 2]. Worldwide, approximately 422 million people suffer from DM, and it is one of the most common non-epidemic causes of physical impairment and mortality [3, 4]. Hyperglycemia, as a result of DM, affects the vasculature of various organs such as the heart, kidneys, nerves and eyes. It induces myocardial infarction, diabetic nephropathy, neuropathy, retinopathy and atherosclerosis [5]. These complications lead to further secondary complications and are frequently associated with delayed wound healing and lower-limb ulceration, which commonly lead to amputation.

Hyperglycemia elevates the levels of unstable reactive molecules, better known as free radicals, that interact with biological molecules thereby increasing the peroxidation of carbohydrates, proteins and lipids and ultimately oxidative stress, leading to an exacerbation of diabetic complications. Free radicals are detoxified and rendered harmless by antioxidant defense systems, thus antioxidants are required to fight against oxidative stress and to prevent biological systems from damage caused by free radicals. It would appear that both endogenous and exogenous antioxidants are essential in avoiding pathophysiological compli-cations caused by DM. Diabetes is the main cause for imbalances between reactive species and antioxidants in the biological system. Due to stress, age, genetic factors, immunodeficiency and various cell signaling abnormalities, oxidative stress is further increased. These factors interconnect with each other creating an environment that promotes the pathogenesis of various diseased conditions [6].

2. OXIDATIVE STRESS AND FREE RADICALS

Regulation of the redox state is critical for normal cellular functioning, and aerobic organisms have mechanisms in place by way of antioxidant systems to block and prevent the harmful effects of oxidants. Oxidative stress occurs when there is an imbalance between oxidants and antioxidants. The increased production of oxidant radicals over antioxidants plays a significant role in the progression of DM and its associated complications. These unstable free radicals are capable of damaging biological molecules resulting in glycoxidation, which is the oxidation of sugars, glycoproteins and glycolipids, and DNA hydroxylation [7, 8]. Free radical induced oxidative stress and its complications in the human body are shown in Fig. (**1**).

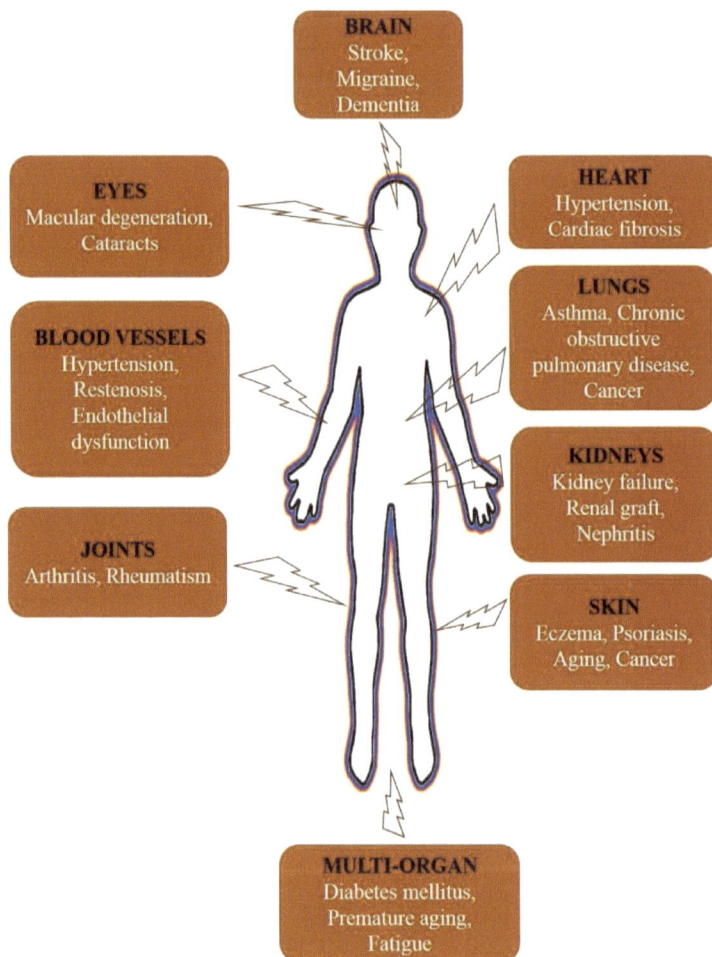

Fig. (1). Free radical induced oxidative stress and its complications in the human body. All most all the organs are affected by oxidative stress; the prolonged oxidative stress ultimately causes various pathophysiological conditions.

Increased free radical formation directly stimulates the immune system, resulting in elevated pro-inflammatory cytokine levels and increased leukocyte infiltration. Increased inflammation and matrix metalloproteinase (MMP) provoke ageing, neurodegenerative disorders, polynephritis and autoimmune disorders. Many metals such as iron (Fe), copper (Cu), cobalt (Co) and nickel (Ni) are oxidized by reactive superoxide anions (O_2^-), resulting in the generation of toxic hydroxyl ions ($^{\cdot}OH$). These types of metallic ions boost free radical generation and augment its interaction with other oxidants *e.g.*, hypochlorous acid (HClO), hypobromous acid (HOBr), chloramines and bromines, leading to the production of more free radicals that affect various tissues and result in visual damage; glaucomatous injury and disease development have been connected with oxidative stress [9, 10].

2.1. Free Radicals

Free radicals are a group of unstable molecules comprised of unpaired electrons which are typically short-lived and reactive; they quickly react with other molecules by either accepting an electron or donating an electron in an attempt to balance their unpaired electrons. The most oxygen-containing free radicals in many diseased states include hydroxyl radicals ($^{\cdot}OH$), superoxide anions (O_2^{-}), hydrogen peroxide (H_2O_2), singlet oxygen ($^1O_2$), hypochlorite (ClO^{-}), nitric oxide (NO) and peroxynitrite ($ONOO^{-}$) [11] Fig. (**2**).

Some researchers do not consider H_2O_2 as a radical as it does not have any unpaired electrons. ROS and RNS are the most important class of free radical molecules produced inside a cell. In biological systems, free radicals are produced during enzymatic reactions for example during phagocytosis and in the mitochondrial electron transport chain, as well as in non-enzymatic reactions involving oxygen with organic compounds and during ionizing reactions [12]. During normal cellular metabolism small quantities of both ROS and RNS are continuously generated. Depending upon the concentration of generated radicals, they function in a positive or negative manner in various tissues [11, 12].

2.1.1. Reactive Oxygen Species (ROS)

ROS are an endogenous source of oxygen free radicals, and under normal physiological conditions they are produced in small quantities during metabolic processes. ROS are generated in various organelles such as the mitochondria, peroxisomes and endoplasmic reticulum. They are chiefly generated during the production of adenosine triphosphate (ATP), fatty acid oxidation and detoxification of xenobiotic processes [13, 14]. ROS are produced as necessary intermediates in picomolar concentrations, and are involved as secondary messengers in cellular signaling, as well as in cell proliferation, differentiation, migration and apoptosis [15]. Above normal physiological levels ROS lead to oxidative stress and damage various biological molecules such as DNA, RNA, proteins and lipids, leading to diseased states [16]. Superoxide anions, hydroxyl radicals and hydrogen peroxide are examples of ROS of the most physiological significance.

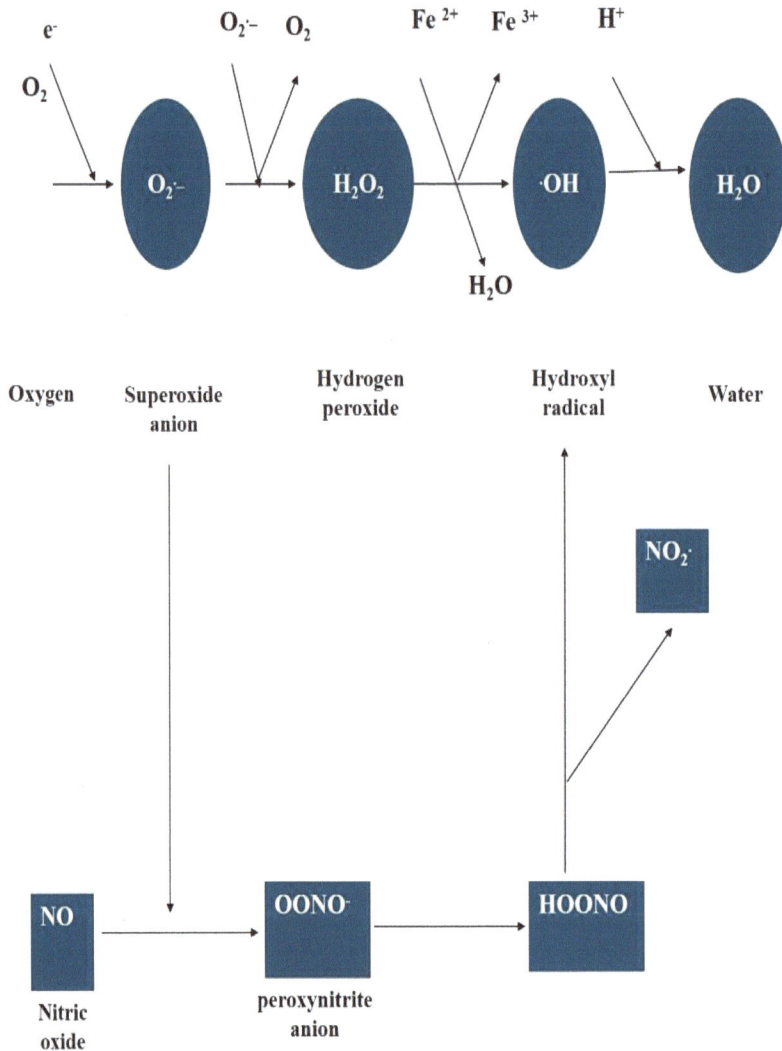

Fig. (2). Generation of free radicals and reactive species. Cells normally generate energy (Adenosine triphosphate, ATP) aerobically by reducing molecular oxygen (O_2) to water (H_2O). The initial transfer of one electron to oxygen leads to sequential redox reactions resulting in the production of multiple reactive oxygen species through the superoxide anion radical, O2•-. The superoxide radical is oxidized to hydrogen peroxide (H_2O_2). Further addition of electrons requires cleavage of the bond between the oxygen atoms, a reaction catalyzed by iron (Fe) (Fenton Reaction) to form the highly reactive hydroxyl radical (OH•). OH• reacts spontaneously with any molecule (protein/carbohydrate/lipids) from which it can abstract a hydrogen atom.

ROS are generated as a by-product during reduction–oxidation (redox) reactions, whereby molecular oxygen (O_2) is reduced (gains an electron) to superoxide anions, the precursor for other reactive species [15]. The reaction is mediated by nicotine adenine dinucleotide phosphate (NAD(P)H) oxidase or xanthine oxidase. Xanthine oxidase catalyses a series of hydroxylation reactions from hypoxanthine to xanthine to yield uric acid. This enzymatic reaction results in the huge production of free radical species [17]. The majority of superoxide anions are produced during ATP production by the mitochondrial electron transport chain. It is mediated by nicotinamide adenine dinucleotide phosphate (NADPH), and molecular oxygen is completely reduced to water. During this process, approximately 1 to 3% of all electrons leak from the system, leading to the production of superoxides [18]. Hydrogen peroxide is generated through the reduction of superoxide anions by the action of superoxide dismutase (SOD) in a process known as dismutation. Hydrogen peroxide is able to diffuse into the cytoplasm where it is embroiled in cellular signaling and regulates protein kinases that are involved in cellular migration and proliferation. It is also capable of diffusing across the cellular membrane and fighting off microbes [19]. Hydrogen peroxide can be broken down into a hydroxyl radical through Fenton reactions, which involves transmission metals such as Fe^{2+} or Cu^{2+}. Hydroxyl radicals are the most reactive species and immediately removes electrons from other molecules, turning them into new radicals that further propagates the formation of ROS [15, 18]. In the presence of chloride ions, hydrogen peroxide can be converted in to hypochlorous acid, which is capable of interacting with DNA and producing pyrimidine oxidation products.

ROS play an important role in the organization of normal wound healing responses. ROS act as secondary messengers to recruit immunocytes and non-lymphoid cells to the wound site, and have the ability to regulate angiogenesis. ROS are involved in host immune responses through phagocytosis, notably by neutrophils, whereby ROS outbursts lead to pathogen destruction, and excess ROS leakage into the surrounding environment has further bacteriostatic effects [20]. ROS also aid in re-epithelialization, whereby hydrogen peroxide triggers the release of essential growth factors and indirectly supports the migration and proliferation of epidermal cells [21].

It is well known that mitochondria function as the major powerhouse of a cell and are involved in ATP production. During this process free radical species are constantly produced within the cell, and a rise in their concentrations is the possible culprits in developing various pathophysiological conditions, including autoimmune disorders, type II diabetes and cardiovascular complications. Many signaling proteins are involved in the formation of mitochondrial ROS. One such protein is p66Shc (p66), which is involved in the regulation of antioxidant enzymes. p66 is encoded for by the ShcA gene and is present in three isoforms as a 46, 52 and 66 kDa protein in mammals. Recent studies have revealed that p66 plays a major role in redox balance regulation. Increased expression of p66 resulted in the decreased expression of antioxidant enzymes by inhibiting FOXO transcription factors, thus elevating oxidative stress [22, 23]. Paneni *et al.* found that the relocation of carbons and electrons away from the mitochondria and towards the cytosol resulted in the reduced generation of mitochondrial superoxide radicals [24].

Mitochondria are known to produce significant amounts of hydrogen peroxide. The production of mitochondrial superoxide ions plays a chief role in normal physiological intracellular signaling. In diseased conditions or treatments (*e.g.*, use of anthracyclins), elevated mitochondrial ROS and oxidative stress result in added lipid peroxidation and subsequent cell membrane and DNA damage, thus stimulating the activation of various protein kinases and signaling cascades that exacerbate the pathophysiological condition further. Any change in mitochondrial fusion and fission processes influences mitochondrial ROS production. Mostly, it generates hydrogen peroxide from the oxidation of endogenous monoamines by the enzyme monoamine oxidase (MAO) located in the mitochondrial outer membrane. In the case of DM, monoamine oxidation occurs at a high rate, releasing large amounts of hydrogen peroxide, thereby increasing oxidative stress that affects various tissues [25].

2.2.2. Reactive Nitrogen Species (RNS)

Under normal physiological conditions L-arginine and oxygen are converted into citrulline and NO through the catalytic activity of nitric oxide synthase (NOS). NO is a highly reactive free radical, with one unpaired electron and a half-life of 15 seconds. The biological activity of newly synthesized NO is promptly bought to an end when it is rapidly oxidized into inorganic nitrite (NO_2^-) and nitrate (NO_3^-), thus completing the NO cycle [26]. Peroxynitrite ($ONOO^-$), a short-lived potent peroxide, is produced through a diffusion-controlled reaction of NO and superoxide radicals. Both ROS and RNS play important biological roles and possess signaling functions in physiological and pathological processes. During wound healing, endothelium-derived NO plays a role in vasodilation and cell signaling, and possesses anti-inflammatory properties.

Nitrosative stress is as a result of an imbalance between RNS and antioxidants. NO is a key marker for endothelial dysfunction and elevated levels may be as a result of an increase in endothelial nitric oxide synthase (eNOS) levels [27, 28]. The activity of numerous enzymes is inhibited by NO, including xanthine oxide, gluthathione peroxidase (GPx), cytochrome c oxidase (COX) and NADPH oxidase (NOX). It interacts with proteins by binding to iron. It does this by binding to a heme group or an iron sulfur complex in enzymes, and in doing so, it either activates or deactivates enzymes [29]. Peroxynitrite is generated in several inflammatory and pathological conditions, and is able to affect a variety of biological molecules by directly attacking sulfhydryl groups in target molecules. NO synthesis may be blocked by inhibiting NOS active sites with asymmetric dimethylarginine (ADMA), a naturally occurring analogue of L-arginine found in plasma and various tissues. Raised levels of ADMA impair endothelial function through the inhibition of NOS [30]. Increased ADMA levels also promote the development of diabetic complications and obesity, and increased arginine methylation plays a significant role in developing cardiovascular disease [31]. Type II diabetes is characterized by elevated serum ADMA levels that affects vascular dysfunction and insulin resistance by elevating oxidative stress [32, 33].

2.2.3. ROS Interaction with Thiol Groups

Low molecular weight (LMW) thiols play a major role in regulating redox-oxidation reactions in cells. LMW thiols are ubiquitous tripeptide glutathione molecules (composed of three amino acids such as Glutamate, Cysteine and Glycine) present in high concentrations in most prokaryotic and eukaryotic cells. In very few prokaryotic species (bacteria and fungi) these LMW thiols are made up of a sulfur molecule with or without glutathione. Some examples of organism-specific LMW thiols include mycothiol (Actinomycetes), bacillithiol (Firmicutes), γ-Glu-Cys (halobacteria and lactic acid bacteria), trypanothione (trypanosomes), ergothioneine (fungi and mycobacteria), Coenzyme M and Coenzyme B (methanogenic archaea) [34]. Most of these LMW thiols are present in high concentrations in combination with disulfides and play an important role in cellular redox reactions. Similar to glutathione, LMW thiols are involved in the formation of disulfide bonds present in proteins that control various metabolic functions [35].

Thiol groups hold a strong interaction with electrophiles, oxidants and metals, making them an important molecule that alters normal cellular functions [36]. Any addition and/or modification in the side chain of thiol groups will determine the positive or negative affect of the thiol [37]. Oxidation of thiols in biological systems (Cys oxidation, sulfenic acids, S-nitrosothiols and disulfides) results in the production of various reversible and irreversible by-products that activate various cellular functions (redox cycling and/or regulation of enzymes) *e.g.* glutathione and thioredoxin effectively cleave free radicals [38].

3. ANTIOXIDANT DEFENSE SYSTEMS

Nature has endowed cells with protective antioxidant defenses to deal with the damaging effects of oxidative stress. Antioxidants are molecules and/or enzymes that are involved in scavenging free radicals and reactive ions; they are the first line of defense against oxidative and nitrosative stress, and minimize undesirable cellular damage caused by ROS/RNS. Antioxidants are defined as substances that prevent substrate oxidation and play a key role in attenuating oxidative stress [39]. Antioxidant enzymes such as SOD, catalase (CAT), GPx, glutathione reductase (GR), glucose-6-phosphate dehydrogenase (G6PD) and thioredoxin reductase (TR), and non-enzymatic antioxidants like glutathione (GSH), uric acid, ceruloplasmin, albumin, bilirubin, NADPH, Coenzyme Q and transferrin, are all endogenous antioxidants [40]. These antioxidants transform ROS to more stable molecules, water and oxygen [11]. Exogenous antioxidants such as carotenoids, phenolic acids, flavonols, thioredoxin, vitamins E and C, and trace metals such as selenium (Se) and zinc (Zn) also function as direct scavengers of ROS [11, 39]. Table **1** depicts the major antioxidant enzymes, as well as endogenous and exogenous antioxidants in biological systems.

Table 1. List of major antioxidant enzymes, and endogenous and exogenous antioxidants found in biological systems. The major enzymatic antioxidant enzymes are superoxide dismutases (EC 1.15.1.1), catalase (EC 1.11.1.6), glutathione peroxidase (EC 1.11.1.19), glutathione reductase, thioredoxin reductase (EC 1.8.1.9), glucose-6-phosphate dehydrogenase (EC 1.1.1.49), and heme oxygenase. Non-enzymatic antioxidants like glutathione, ceruloplasmin, albumin, bilirubin, uric acid, lipoic acid, transferrin, NADPH, Coenzyme Q, flavonoids, quercetin, curcumin, vitamin C and E, ascorbic acid, and selenium also play a crucial role in scavenging free radicals and ROS due to its capability to donate protons to neutralize ROS, thereby preventing oxidative stress.

Enzymatic Antioxidants	Non-Enzymatic Antioxidants	Exogenous Antioxidants
Superoxide dismutase (SOD)	Glutathione (GSH)	Flavonoids, Lycopene
Catalase (CAT)	Ceruloplasmin	Quercetin, Curcumin
Glutathione peroxidase (GpX)	Albumin, bilirubin	Vitamin C & E
Glutathione reductase (GR)	Uric acid, Lipoic acid	Alpha-tocopherol
Glucose-6-phosphate dehydrogenase (G6PD)	NADPH, Coenzyme Q	Ascorbic acid
Heme oxygenase (HMOX1)	Transferrin	Selenium
Thioredoxin reductase (GR)		

Endogenous antioxidant defences are incomplete without exogenous antioxidants, and there is an incessant need for exogenous antioxidants to prevent oxidative stress. Antioxidants such as vitamin C, vitamin E and vitamin A are not produced by the body and must be taken exogenously through diet or supplements. High doses of isolated exogenous antioxidants may be toxic, which can be due to pro-oxidative effects at high concentrations or their ability to react with beneficial, normal physiological concentrations of ROS necessary for optimal cellular functioning [39, 40]. Apart from possessing free radical scavenging properties, antioxidants are also involved in regulating various signaling pathways related to cell survival, for example catechin a well-known exogenous antioxidant is involved in modulating activator protein-1 (AP-1) and nuclear factor kappa B (NFkB) in endothelial cells [41]. An overview of the cellular antioxidant defense system is explained in Fig. (**3**).

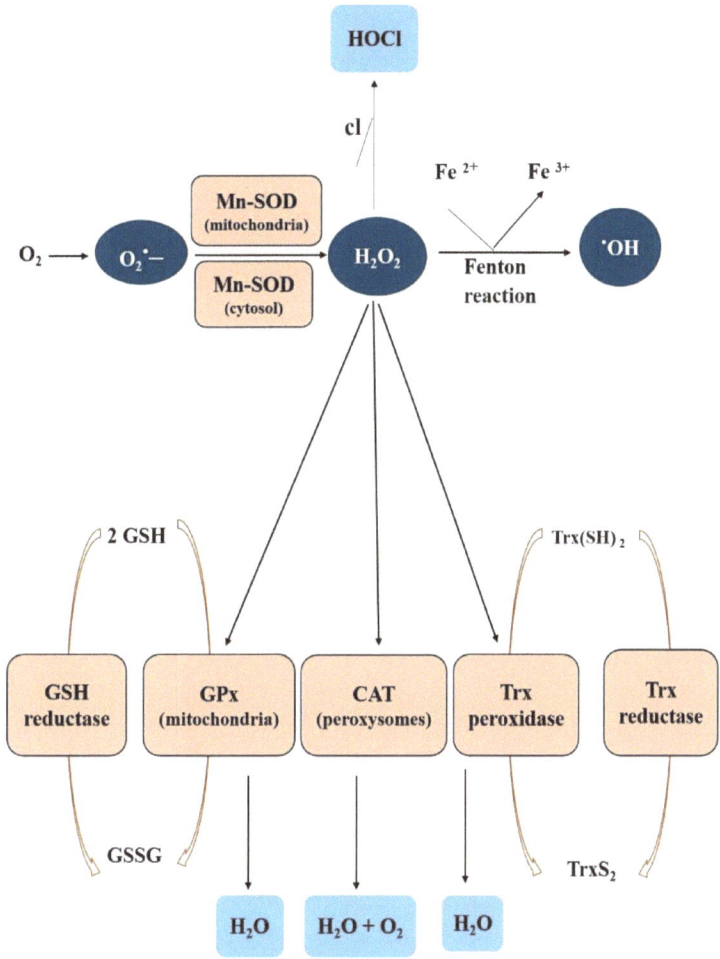

Fig. (3). Cellular antioxidant defense systems. Cellular antioxidant defense systems works synergistically with each other (enzymatic and non-enzymatic antioxidants) to protect the cells and organ systems of the body against oxidative stress. The major enzymes such SOD, CAT and GPx effectively eliminate most of the reactive oxygen/nitrogen species. Mn-SOD- mitochondrial superoxide dismutase, CAT- catalase, GPx- glutathione peroxidase, GSH reductase- glutathione reductase, Trx peroxidase- thioredoxin peroxidase, Trx reductase- thioredoxin reductase, HOCL- hypochlorous acid.

Many naturally occurring antioxidants are obtained from phytochemicals, also referred to as phytonutrients, which are present in plants [42]. Phytochemicals include compounds such as flavonoids, betacarotene, chlorophyll and antho-cyanins. Although phytochemicals are biologically active, they are not nutritive. Due to their high antioxidant content, many plants are used for their medicinal properties to treat diabetes and associated complications. Several lines of evidence suggest that plants, plant extracts and isolated plant compounds have less cellular toxicity to humans than chemical drugs for the treatment of DM [43, 44].

3.1. Antioxidant Mechanisms Against Free Radicals

SOD, CAT, GPx and GR are antioxidant enzymes which play a fundamental and indispensable role in the antioxidant protective capacity of biological systems against free radical attack. The Fenton reaction occurs chiefly in peroxisomes, which is strictly controlled by CAT, and involves the breaking down of hydrogen peroxide to yield water and oxygen, thereby continuously restricting free radical-induced tissue damage. Since CAT is only present in peroxisomes, there is an accumulation of hydrogen peroxide in the mitochondria. This hydrogen peroxide is reduced to water and peroxides by GPx, whereby these peroxides are then further broken-down to their respective alcohol byproducts. This type of enzymatic reaction, which involves the breakdown and elimination of reactive species, is considered as the first line of antioxidant defense. Free radical scavenging enzymes such as SOD, CAT, GPx and GR are crucial for defenses against oxidative stress, particularly in response to super oxide anion radical ($O2^{-}$) formation, which is continuously produced during cellular metabolism.

3.2. Superoxide Dismutase (SOD, EC 1.15.1.1)

SOD is the main endogenous antioxidant enzyme generated within cells, and acts as the first defense against reactive species and free radicals. SOD catalyzes the cleavage of superoxide anions ($^{-}O_2$) to hydrogen peroxide (H_2O_2) and molecular oxygen (O_2), subsequently removing the harmful effects of superoxide anions. SOD is a metalloenzyme and hence requires a metal cofactor for its activity. Based on the type of metal ion cofactor required by SOD, various forms of the enzyme exist [44, 45]. Metals such as iron (Fe), zinc (Zn), copper (Cu) and manganese (Mn) serve as cofactors, and based on this SODs are known as Cu/Zn-SOD, Mn-SOD, and Fe-SOD. Cu/Zn-SOD is usually found in eukaryotes where it is located mostly in the cytosol and peroxisomes, and is encoded for by SOD1 located on chromosome 21. Mn-SOD is normally present in both prokaryotes and eukaryotes, and is encoded for by SOD2 located on chromosome 6. Fe-SOD is normally present in prokaryotes and chloroplasts of some plants, and is encoded for by SOD3 located on chromosome 4 [46, 47]. SOD3 plays a potential role in antioxidant defense against ischemic reperfusion injury, as well as cardiovascular and neurological diseases. Lebovitz and colleagues found that SOD knock out mice displayed evidence of neuronal damage, myocardial infarction and perinatal mortality [48]. Recently, Dayal and colleagues found that in hyperhomocysteinemia induced mice, lower levels of SOD promoted cerebral vascular hypertrophy and vascular dysfunction [49].

3.3. Catalase (CAT, EC 1.11.1.6)

CAT is an enzymic antioxidant present in all living tissues that consume oxygen. CAT is a 240 kDa tetrameric protein composed of four subunits. It contains a ferriprotoporphyrin structure and is encoded by *cttl* gene mapping to chromosome 11 [45, 50]. CAT is activated once it is bound to the cofactor iron (Fe) or manganese (Mn) and reduces hydrogen peroxide to water and oxygen, subsequently finishing the detoxification process initiated by SOD [51, 52]. CAT is highly expressed inside cells, where it constantly scavenges hydrogen peroxide within one second of its generation. CAT also reacts well with other hydrogen donors like methanol, ethanol, formic acid and phenols. CAT is activated during a two-step reaction; i) at first heme is oxidized to an oxyferryl species by hydrogen peroxide. During this process, when the porphyrin ring in iron gets oxidized, it generates cation radicals that subsequently lead to the formation of other free radicals. ii) Hydrogen peroxide acts as a strong reducing agent to stimulate the inactive enzyme (CAT) and promotes the formation of oxygen and water [52]. At low concentrations, hydrogen peroxide regulates various physiological processes such as cell proliferation, cell cycle, carbohydrate metabolism, mitochondrial function and platelet activation, but is dangerous to cells when it reaches higher concentrations [53]. Therefore, the ability of CAT to efficiently lower hydrogen peroxide concentrations emphasizes its significance in various physiological processes, as well as being a first line antioxidant defense enzyme.

3.4. Glutathione Peroxidases (GPx, EC 1.11.1.9)

GPx is another important enzymic antioxidant that is involved in the breakdown of hydrogen peroxide to water, and lipid peroxides to their corresponding alcohols. GPx is mostly active inside the mitochondria, and in some cases in the cytosol [54]. GPx activity mainly depends on the cofactor selenium (Se), and it is commonly known as selenocysteine peroxidase. The major function of GPx is to attenuate the peroxidation of lipids, thereby protecting cells from cell damage induced cell death. In humans, a minimum of eight different forms of GPx are present. Genes for GPx1 to GPx8 are mapped to chromosomes 3, 14, 5, 19, 6, 6, 1 and 5, respectively. Among the various forms of GPx, the most common is GPx1 and GPx7 that is located in nearly all cell types. GPx2 is located only in the gastrointestinal tract, whereas GPx3 can be found in the kidneys. GPx4 is found in fat tissues; GPx5 is found in the mammalian male reproductive tract; GPx6 is found in the olfactory lobe; and GPx8 is found in the putative gland [55, 56]. Except for GPx4, all other forms of GPx are tetrameric in structure. GPx4 is made up of a phospholipid hydroperoxide monomer and possesses different cofactor receptors; hence GPx is involved in the breakdown of phospholipid hydro-peroxides [57]. GPx5 is present in both humans and rodents, whereas GPx6 is only found in humans and differs from all other GPx forms in that their functions are independent of selenium [45, 56]. Chabory *et al.*, found that diabetic patients with reduced GPx levels resulted in impaired antioxidant production, which results in increased oxidative stress leading to neuronal damage [58]. Forgione and colleagues found that lower levels of GPx1 promote oxidative stress in the blood vessels, leading to endothelial dysfunction and cardiovascular disease [59].

3.5. Antioxidant Vitamins

Some vitamins function as antioxidants by quenching free radicals and maintaining redox balance. In this aspect, α-Lipoic acid (ALA) and vitamin C and vitamin E function as antioxidants. Various experimental and clinical studies have found that vitamin antioxidants and vitamin supplements can significantly reduce reactive species, free radicals and lipid peroxidation, and in doing so diminish oxidative stress. ALA is well known for its potential role against oxidative stress. In DM it effectively removes free radicals, thereby enhancing cellular glucose uptake and reducing diabetic complications. Maritim and colleagues observed that treatment of diabetic rats with ALA significantly reduced free radicals and oxidative stress by lessening lipid peroxidation [60]. Data demonstrates that in diabetic patients, the levels of non-enzymatic antioxidants such as beta-carotene and vitamin C and E are low, resulting in increased oxidative damage [61, 62]. Recent research suggests that vitamin C is involved in the generation of NO in epithelial cells. In diabetic patients, the risk of cardiovascular disease is high and the possible reason is that the high glucose environment increases NO levels, which migrate towards smooth muscles thereby affecting the functioning of various organs such as the heart and kidneys [61].

Vitamin C and E are the most important and most studied antioxidant vitamins. Vitamin C is a hydrophilic molecule which effectively removes radical species, mostly hydroxyl radicals. Vitamin E is a fat-soluble lipophilic molecule that possesses potent free radical scavenging activity (mainly detoxify free radicals that cause lipid peroxidation). Vitamin E is present in eight different fat-soluble forms. α- tocopherol is the major active form of vitamin E that is found in sunflowers, safflowers and wheat germ oil. Both vitamin C and E act in a synergistic manner; at first vitamin E is oxidized to tocopheroxyl radicals and further reduced to tocopherol by vitamin C and GSH. Vitamin C is involved in the regeneration of vitamin E through a reduction-oxidation cycle, and it elevates the levels of intracellular glutathione. Like vitamin A, a combination of vitamin C and E can be used at high doses to help avoid diabetes and cardiovascular disease [63].

3.5.1. Phytochemicals

Many reports confirm that plants and plant-derived compounds possess strong antioxidant activity, such as betacarotene, diterpenes, allicin, hesperidin, lycopene, phenylpropanoids, tocopherols, flavonoids, resveratrol, quercetin, tannins, curcumin, cathechin, caspsanthin, cinnamic acids and triterpenes [64, 65]. Any damage to mammalian cells leads to lipoxygenase activation that catalyzes the formation of hydroperoxide radicals; these radicals might react with fatty acids to generate dioxolanes that are responsible for antioxidant defenses. Therefore, damage caused by oxidative stress might be protected against or resolved using plants, predominantly with high levels of active antioxidants [66, 67]. Polyphenols are chemical compounds found in plants, which are composed of aromatic ring(s) bearing one or more hydroxyl moieties. They exhibit antioxidant behavior and are divided into four classes including stilbenes, lignans, flavonoids and phenolic acids [68]. Polyphenolic compounds inhibit oxidation through various mechanisms. One such mechanism involves the delocalization of the gained electron over the phenolic antioxidant and stabilization of the free radical by the resonance effect of the aromatic nucleus [69].

4. OXIDATIVE STRESS AND DIABETES

Various researchers have extensively studied micronutrient and antioxidant ailments in DM, and there is evidence to prove that oxidative stress is the culprit in the development of diabetic complications [70, 71]. Though type I diabetes is an autoimmune disorder, chemical agents such as alloxan and streptozotocin will result in insulin deficiency by damaging pancreatic β-cells, which can be overcome with the use of antioxidants [71].

Antioxidant defense systems are an essential function for cell survival, and oxidative stress is as a result of the imbalance between the increased generation and\or reduced clearance of free radical species [72]. A decrease in antioxidant defense systems results in increased oxidative stress that has deleterious effects on cellular glucose uptake and metabolism. The increased production of various types of free radicals such as superoxide anions and hydrogen peroxide could inactivate the enzymes involved in glycolysis (glucokinase, phosphofructokinase and pyruvate kinase), resulting in hyperglycemia [73]. It has been reported that increased oxidative stress results in increased calcium uptake and depolarization by pancreatic β-cells, creating hyperinsulinemia (increased insulin resistance and downregulation of insulin receptors) [74, 75]. Moreover, increased blood glucose levels stimulate various signaling pathways which increase the levels of glucose oxidation, peroxides and its by-products, adding to oxidative stress. There is also an increase in fructose levels, and the enzymatic reactions involved in fructose degradation *via* aldose-reductase further promote the generation of free radicals [76]. Studies suggest that the enzyme protein kinase C (PKC) plays a crucial role in the development of diabetic complications. PKC is involved in modulating the function of other proteins through phosphorylation reactions. In diabetes, PKC functions directly to stimulate the production of eNOS and NO, thereby it increases protein degradation and endothelial dysfunction [77, 78]. It is possible that activation of PKC acts as a key signal for various cellular pathways that exacerbate the diabetic condition. Elevated carbonyl damage in diabetes results in increased oxidation of glucose and lipids, thereby promoting prolonged oxidative stress and tissue damage [70, 71].

Malondialdehyde (MDA), a toxic molecule regarded as a common marker of oxidative stress, is an example of an end product from oxidation reactions involving lipids [79]. Bhutia *et al.,* conducted a study in type II diabetic patients where they assessed MDA levels. It was found that MDA levels were significantly elevated in diabetic patients compared to non-diabetic patients. Increased free radicals in DM leads to the amplified peroxidative breakdown of phospholipids resulting in increased MDA levels, which in turn react with deoxyguanosine and deoxyadenosine in DNA, producing DNA adducts. They also play a significant role in the modification of low density lipoproteins (LDL) that mediate pathophysiological changes through non-enzymatic and auto-oxidative glycosylation [80]. The Amadori rearrangement (rearrangement reaction of N-glycoside in aldose or glycosylamine to the corresponding 1-amino-1-de-xy-ketose) of Schiff base products, as observed with glycated hemoglobin, is an intermediate product in the production of advanced glycation end products (AGEs). Glycated hemoglobin levels are used as a marker for diabetes, and increased levels are found in diabetic states [81]. ROS interact with proteins and cause several reactions that lead to peptide cleavage and amino acid oxidation, resulting in protein damage. Oxidation of DNA results in DNA strand damage, and allows ROS to interact with micro-RNA that leads to irregular protein translations [82]. The implications of hyperglycemia in elevating oxidative stress are outlined in Fig. (**4**).

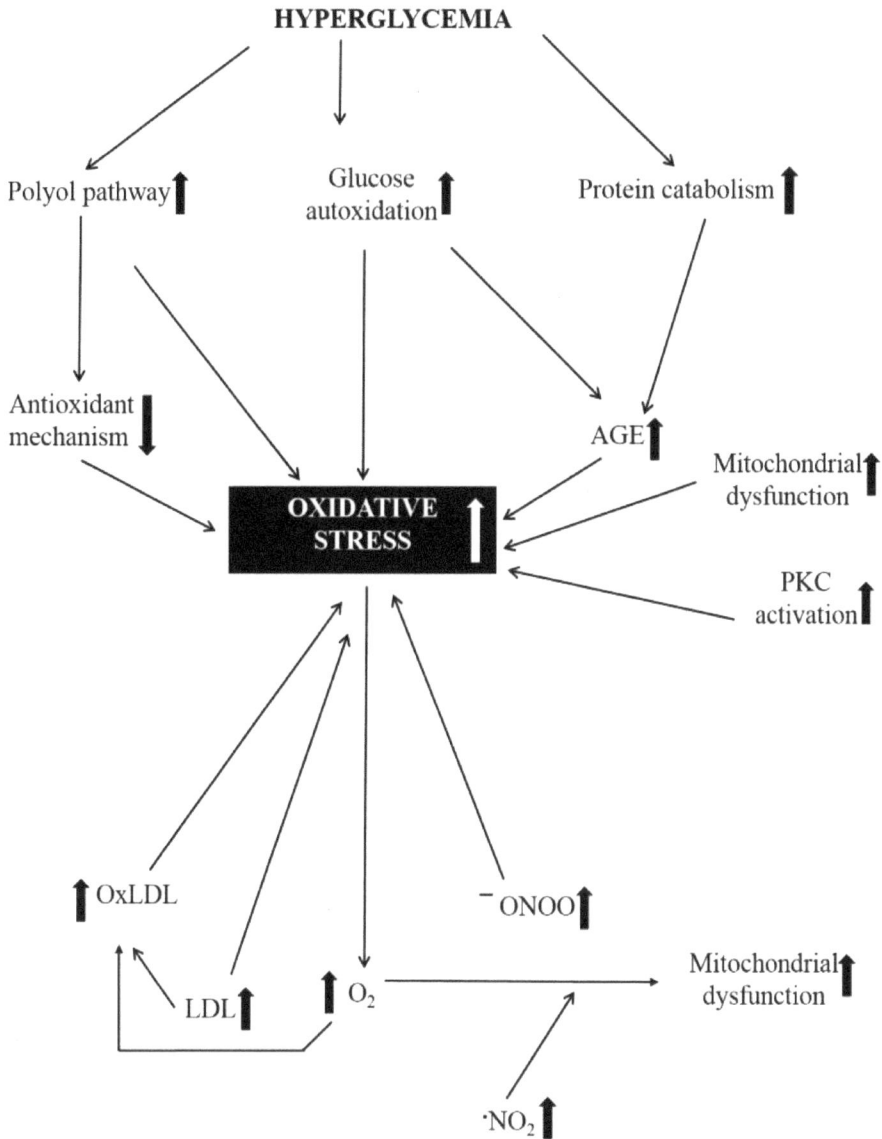

Fig. (4). Implications of hyperglycemia in elevating oxidative stress. A diabetic environment increases the production of reactive oxygen species (ROS) and induces many metabolic abnormalities, including activation of the polyol pathway, protein catabolism, advanced glycation end products, mitochondrial dysfunction and production of inflammatory mediators. Increased ROS damages mitochondria and increases the oxidation of proteins and lipids. AGE- advanced glycation end products; PKC- protein kinase C; .NO$_2$- nitrogen dioxide; O$_2$- singlet oxygen; LDL- low density lipoprotein; ox-LDL- oxidized low density lipoprotein; $^-$ONOO-peroxynitrite.

4.1. Vascular Inflammation in Diabetes

Diabetic retinopathy is one of the most important causes of impaired vision in adults. The reason for this complication is mostly attributed to microvascular damage, swelling of the blood vessels and fluid leaks. If diabetic retinopathy is not prevented, new blood vessels grow and proliferate on the surface of the retina and finally leads to retinal detachment [83]. The occurrence of this disease depends on the duration of diabetes, and rarely develops in the early stages of diabetes. However, the rate increases to 50% at 10 years and 90% at 25 years of having DM. The major reason for the development of complications in diabetic retinopathy is prolonged oxidative stress induced by uncontrolled blood sugar levels and an imbalance in redox homeostasis. Oxidative stress is not only involved in the development of diabetic retinopathy, but also in impaired wound healing, even after the restoration of glucose levels. Diabetic retinopathy is drug resistant and the damage caused by diabetes cannot be restored due to increased lipid peroxidation and oxidative damage in the retina due to the accumulation of free radicals [84, 85]. An increase in oxygen consumption and prolonged contact to light induces oxidative stress and directly damages the retina. Several studies revealed that in diabetic retinopathy, photochemical damage is the most common cause of elevated oxidative stress, and these molecules are encountered by various enzymic and vitamin antioxidants [86, 87]. Hence, uncovering markers of oxidative stress for early detection of diabetic retinopathy and the intake of antioxidant supplements could be used in the prevention of retinopathy in diabetic patients.

Endothelium and endothelial cells play a major role in regulating vascular homeostasis (vasodilation/vasoconstriction, blood clot formation, and activation of platelets and leukocytes). The endothelium regulates the vascular environment by producing/releasing various vasodilators and vasoconstrictors, *e.g.*, NO, bradykinin, potassium ions, adenosine and endothelin [88]. Vascular health depends on the normal functioning of endothelial cells (ECs). EC dysfunction results in the over expression of cell adhesion molecules that attract inflammatory cells and damage vascular tissue. Any dysfunction or abnormality in ECs damage vascular tissue. For example, during the atherothrombotic process, EC dysfunction is one of the initial stages, which contribute to the progression of vascular blockage and damage. Other than EC dysfunction, systemic inflammation also plays a major role in promoting vascular damage. In diabetic patients, both EC dysfunction and systemic inflammation are connected with prothrombotic and hypofibrinolytic factors, which promotes vascular occlusion leading to the formation of a myocardial infarction, stroke, and peripheral vascular disease. Under diabetic conditions, insulin resistance increases blood glucose levels that in turn disturb vascular homeostasis, leading to endothelial dysfunction and the progression of atherosclerotic conditions [89].

During diabetes, increased ROS and reduced NO bioavailability contribute to vascular disease progression. Insulin receptors play a major role in maintaining normal vascular homeostasis; under normal physiological conditions, insulin receptors inhibit the production of NO by blocking the activity of eNOS that results in normal vasodilation [90]. In diabetic patients, the number of insulin receptors is reduced resulting in insulin resistance and the reduction in vasodilators, and increased production of vasoconstrictors. Endothelin-1 is a well known vasoconstrictor and their levels are increased in diabetic patients, leading to endothelial dysfunction and progression of microangiopathic conditions [91]. Platelet activation is related to the bioavailability of NO. In a study conducted in diabetic mice, the inhibition of NOS reduces the phosphorylation of vasodilator stimulated phosphoprotein (VASP), and increases the binding effect of fibrinogen-platelets and P-selectin levels, as well as increasing the expression of CD40. These results show the importance of NO production by ECs in regulating platelet activation to maintain normal vascular health [92].

In diabetes, both elevated blood glucose and free fatty acid levels increase ROS generation that in turn stimulate NO synthesis by various signaling pathways. More specifically, NF-κB signaling is activated when fatty acids bind to the toll-like receptor. Activated NF-κB stimulates the release of inflammatory cytokines such as interleukin (IL)-6 and tumor necrosis factor (TNF)-α. Other than NF-κB activation, the toll-like receptor promotes insulin receptor substrate-1 phosphorylation by activating c-Jun amino-terminal kinase (JNK) and PKC. This results in the downregulation of the PI3-kinase/Akt pathway. The generation of eNOS and NO can be lowered by suppressing the PI3-kinase/Akt pathway. In various pathophysiological conditions, increased blood glucose levels and increased oxidative stress upregulate the activation of PKC and NF-κB signaling pathways that promotes vascular inflammation and atherosclerosis [92].

4.2. Antioxidant Strategies to Control Oxidative Stress in Diabetes

During overnight fasting, the rate of production and utilization of glucose is equal, stabilizing plasma glucose levels. In healthy adults the overnight fasting plasma glucose level ranges from 1.8 to 2.6 mg/kg-1/min-1 [93]. But in the case of taking a meal, there will be changes in the rate of glucose production/utilization. Glucose absorption will result in increased transport of exogenous glucose into the blood circulation that will be two times more than the rate of postabsorptive endogenous glucose production, and is dependent on the carbohydrates present in the meal and the rate of glucose absorption by individual biological systems. Once exogenous glucose is completely absorbed, there will be a halt in endogenous glucose production that stimulates the utilization of glucose by the liver, muscle and fat cells [93]. Finally, plasma glucose concentrations return to their normal fasting state *i.e.* 1.8 to 2.6 mg/kg-1/min-1. In the case of diabetic patients, pancreatic islet dysfunction, insulin resistance, insulin deficiency and insulin receptor dysfunction is responsible for increased blood glucose levels and poor glucose utilization that contributes to the development of hyperglycemia. Insulin plays a very important role in controlling postprandial glucose in three possible ways. i) Insulin commands cells present in peripheral tissue and muscles to increase their utilization of glucose [94]. ii) Insulin promotes glycogenesis (formation of glycogen from glucose in the liver). iii) Insulin inhibits the production of glucose *via* glycogenolysis and gluconeogenesis by inhibiting the secretion of glucagon in pancreatic β-cells. The main function of insulin is to maintain blood glucose levels.

Increasing evidence proposes that DM is associated with elevated oxidative stress with inadequate antioxidant defense responses [61, 95]. For example, the elevated generation of superoxide radicals contributes to endothelial dysfunction. Guzik and co-workers found that increased sugar levels in diabetic patients increased vascular superoxide, PKC and NAD(P)H oxidases, and reduced the bioavailability of NO [96]. Antioxidants are acquired from various dietary sources and are used for scavenging free radicals to avoid oxidative stress. Increased oxidative stress in diabetes is the major reason for sustained inflammation and endothelial dysfunction. This condition can be overcome with antioxidant treatment, and several studies revealed that the intake of antioxidant rich foods and supplements helps to fight against oxidative stress and avoid diabetic complications. Even though many clinical and experimental studies have revealed a positive role of antioxidants in detoxifying free radicals, very little evidence is available indicating the role of antioxidants in maintaining glucose levels and progression of complications [11, 97]. There is a lack in clinical studies explaining the use and potential role of antioxidants in diabetes: this may be due to the fact that ROS have a very short half-life, and even at low concentrations they activate various biochemical processes, and cell and insulin signaling. Moreover, some antioxidants hold diverse properties other than an antioxidant property, for example, vitamin E functions as a strong antioxidant and it is also involved in the activation of *COX2* and *NFκB* genes [98]. Once antioxidants react with free radicals, they are converted to a pro-oxidant. Later, these pro-oxidants act as an antioxidant to detoxify free radicals preventing tissue damage. There is limited evidence suggesting that antioxidant treatment extends a healthy life of diabetic patients or improves glycaemic control with current management strategies.

In DM, antioxidants play a role in protecting cells against oxidative bursts which in turn helps lowering hyperglycemia. The overexpression of the antioxidant enzyme CAT has therapeutic potential in diabetic-induced cardiomyocyte dysfunction. The study by Turdi *et al.*, showed that CAT had a positive effect on reducing increased levels of ROS, preventing apoptosis and attenuating diabetic induced cell signaling associated with Akt, forkhead transcription factor and silent information regulator 2 [99]. In streptozotocin induced diabetic Wistar rats, the use of antioxidants such as vitamin A, E and C protected the kidneys against oxidative stress induced diabetic nephropathy by reducing the formation of MDA and enhancing antioxidant levels, thus allowing the kidneys to perform their normal function [100].

Many experimental studies confirm the positive role of antioxidants in treating DM. However, clinical data is not sound. In some cases, antioxidant supplementation is not always beneficial as observed in a recent study involving N-acetylcysteine (NAC) in human subjects. NAC is the precursor for GSH and helps in increasing GSH levels. NAC supplementation in patients with type II diabetes had little or no effect on increased blood glucose levels and other oxidative stress markers such as GSH, GSH/oxidized glutathione ratio, thiobarbituric acid reactive substances and urine F2-isoprostanes. Conversely, the study suggested that high doses of NAC could in fact be detrimental by increasing blood glucose levels [44]. Clinical studies have found that supplementation with vitamin E, vitamin C and beta carotene did not show any significant result on the prevention of type II diabetes. Though not statistically significant, there was a slight decrease in the risk for developing DM in women who received vitamin C, whereas patients who received vitamin E showed a slight increase in developing diabetes [101]. Most clinical trials have failed to show the protective effects of antioxidants, with the exception of ALA. ALA is an organosulfur compound derived from caprylic acid and is vital for aerobic metabolism. It acts as a cofactor in multienzyme complexes, including pyruvate dehydrogenase and branched-chain ketoacid dehydrogenase. ALA is important in the treatment of diabetic neuropathy, and is involved in the improvement of insulin sensitivity [100]. A possible reason for clinical trial failure in the use of antioxidants might be due to the population under study, where it was observed that diabetic patients with high oxidative stress did not responded well to vitamin E treatments. Also, a standard measure to compare baseline oxidative stress for all patients is not available, therefore contradicting effects of antioxidant vitamins in diabetes management have been observed [61].

With the discovery of SOD metal complexes, the quest for synthesizing new SOD mimics has gained interest. These complexes have been used in experiments pertaining to human health. Among the different metal complexes, Mn-SOD has been shown to be beneficial over Fe-SOD due to the redox activity of Fe and the risk of producing free radicals through Fenton-reactions resulting in cellular toxicity *in vivo* [102]. Among Mn complexes, Mn(II) cyclic polyamines, Mn(III) porphyrins, Mn(III) salen and manganese peroxidases (MnPs) are being studied for their beneficial role [103]. Despite the fact that they differ in physical and chemical properties, the activity of these mimics is more or less similar; however, Mn(III) salen is considered as the better choice [104]. Hence, the properties of these compounds are expected to overlap when used at the laboratory or clinical level. Clinical efficiency of Mn(II) cyclic polyamines and Mn(III) porphyrins are under way, while Mn(III) salen is already in the market and is used in cosmetics. Two Mn(III) porphyrin analogs MnTnBuOE-2- PyP5+ (BMX-001) and MnTE-2-PyP5+ (AEOL10113, BMX-010) are currently being studied in clinical trials, giving us the understanding that more and more SOD mimics will be available in the future [105].

CONCLUSION

DM and its associated complications might damage various organs such as the heart, eyes, kidneys and nerves. Micro- and macrovascular impairment is the foremost reason for mortality in diabetic patients. High blood glucose levels elevate ROS and generate oxidative stress that plays a major role in the deleterious effect of diabetes. ROS and RNS play important physiological functions and are part of many signaling pathways involved in wound healing, however they can also cause extensive cellular damage. Therefore, there is a need for antioxidants to fight against oxidative stress and counter balance the effects of ROS and RNS. The positive role of antioxidants in scavenging free radicals has been frequently studied. However, simultaneous research findings reveal the dark side of antioxidants which could be detrimental to human health, and as such a balance in the current treatment with antioxidants should exist based on the accumulating evidence of recent research reports. The various aspects of antioxidants, both as a preventive and causative risk factor in various human diseases as explained in this review, suggest that a multitude of factors have to be considered in the prescription of antioxidants as a preventive measure to decrease the risk of certain diseases. Combinatorial therapy using antioxidants is one of the ways in which to harbor the beneficial effects of antioxidants and to reduce the risk of the aggravating disease. Also, tailoring antioxidants to specific locations associated with oxidative outbursts could be potentially useful to treat various diseases associated with oxidative stress. The negative cases observed for a few antioxidants should not be generalized to all antioxidants. Exploration of the mechanism of action and the optimization of concentrations to be administered according to the physiological sites might help in improving the treatment efficacy of antioxidants, making them more of a friend than a foe. The main challenge in diabetic research is to find an appropriate way to attenuate oxidative stress. However, the lack of clinical data to prove the positive effects of vitamin antioxidants in the prevention of various oxidative stress induced disease conditions has led to the development of new therapeutic strategies to scavenge mitochondrial free radicals. On the other hand, it is not clear whether the benefits of antioxidant supplementation depend on standard antioxidant levels, since currently there are no reliable biomarkers of oxidative stress available for large clinical studies. In conclusion, with regard to oxidative stress therapy, it seems important to develop new compounds that can scavenge free radicals and radical generating enzymes, and imitate endogenous antioxidants to prevent diabetes-associated complications.

CONSENT FOR PUBLICATION

Not applicable.

CONFLICT OF INTEREST

The authors confirm that there is no conflict of interest to declare for this publication.

ACKNOWLEDGEMENTS

This work is based on the research supported by the South African Research Chairs Initiative of the Department of Science and Technology and National Research Foundation of South Africa (Grant no 98337); the University of Johannesburg; and the National Research Foundation of South Africa. Funding sources had no involvement in study design, data collection, analysis and interpretation, writing of the report, and in the decision to submit the article for publication.

REFERENCES

[1] Patel VS, Chitra V, Prasanna PL, Krishnaraju V. Hypoglycemic effect of aqueous extract of *Parthenium hysterophorus* L. in normal and alloxan induced diabetic rats. Indian J Pharmacol 2008; 40(4): 183-5.
[http://dx.doi.org/10.4103/0253-7613.43167] [PMID: 20040954]

[2] Chintan AP, Nimish LP, Nayana B, Bhavna M, Mahendra G, Hardik T. Cardiovascular complication of diabetes mellitus. J Appl Pharm Sci 2011; 4: 1-6.

[3] Adeloye D, Ige JO, Aderemi AV, *et al.* Estimating the prevalence, hospitalisation and mortality from type 2 diabetes mellitus in Nigeria: a systematic review and meta-analysis. BMJ Open 2017; 7(5)e015424
[http://dx.doi.org/10.1136/bmjopen-2016-015424] [PMID: 28495817]

[4] 2018.https://www.who.int/news-room/fact-sheets/detail/diabetes

[5] Chawla A, Rajeev C, Jaggi S. Microvascular and macrovascular complications in diabetes mellitus: Distinct or continuum? Ind J Endocrin Met 2016; 67: 157-9.

[6] Bullone M, Lavoie JP. The Contribution of Oxidative Stress and Inflamm-Aging in Human and Equine Asthma. Int J Mol Sci 2017; 18(12): 2612-22.

[http://dx.doi.org/10.3390/ijms18122612] [PMID: 29206130]

[7]   Qutub AA, Popel AS. Reactive oxygen species regulate hypoxia-inducible factor 1alpha differentially in cancer and ischemia. Mol Cell Biol 2008; 28(16): 5106-19.
[http://dx.doi.org/10.1128/MCB.00060-08] [PMID: 18559422]

[8]   Rani V, Deep G, Singh RK, Palle K, Yadav UC. Oxidative stress and metabolic disorders: Pathogenesis and therapeutic strategies. Life Sci 2016; 148: 183-93.
[http://dx.doi.org/10.1016/j.lfs.2016.02.002] [PMID: 26851532]

[9]   Zapata GL. The role of Oxidative Stress in Ocular Disease. Stud on Vet Med 2011; 30: 113-31.
[http://dx.doi.org/10.1007/978-1-61779-071-3_8]

[10]   Lai JY, Luo LJ. Antioxidant gallic acid-functionalized biodegradable in situgelling copolymers for cytoprotective antiglaucoma drug delivery systems. Biomacromolecules 2015; 16(9): 2950-63.
[http://dx.doi.org/10.1021/acs.biomac.5b00854] [PMID: 26248008]

[11]   Rahal A, Kumar A, Singh V, *et al.* Oxidative stress, prooxidants, and antioxidants: the interplay. BioMed Res Int 2014; 2014761264
[http://dx.doi.org/10.1155/2014/761264] [PMID: 24587990]

[12]   Zorov DB, Juhaszova M, Sollott SJ. Mitochondrial reactive oxygen species (ROS) and ROS-induced ROS release. Physiol Rev 2014; 94(3): 909-50.
[http://dx.doi.org/10.1152/physrev.00026.2013] [PMID: 24987008]

[13]   Murphy MP. How mitochondria produce reactive oxygen species. Biochem J 2009; 417(1): 1-13.
[http://dx.doi.org/10.1042/BJ20081386] [PMID: 19061483]

[14]   Sharma P, Jha AB, Dubey RS, Pessarakli M. Reactive Oxygen Species, Oxidative Damage and Antioxidant Defense Mechanism in Plants under stressful conditions. J Bot 2012; 19: 1-26.

[15]   Beckhauser TF, Francis-Oliveira J, De Pasquale R. Reactive oxygen species: physiological and physiopathological effects on synaptic plasticity. J Exp Neurosci 2016; 10 (Suppl. 1): 23-48.
[http://dx.doi.org/10.4137/JEN.S39887] [PMID: 27625575]

[16]   Phaniendra A, Jestadi DB, Periyasamy L. Free radicals: properties, sources, targets, and their implication in various diseases. Indian J Clin Biochem 2015; 30(1): 11-26.
[http://dx.doi.org/10.1007/s12291-014-0446-0] [PMID: 25646037]

[17]   Kostic DA, Dimitrijevic DS, Stojanovic GS, Palic IR, Dordevic AS, Ickovski JD. Xanthine Oxidase: Isolation, Assay of Activity and Inhibition. J Chem 2015; 1-8.
[http://dx.doi.org/10.1155/2015/294858]

[18]   Ponugoti B, Dong G, Graves DT. Role of forkhead transcription factors in diabetes-induced oxidative stress. Exp Diabetes Res 2012; 2012939751
[http://dx.doi.org/10.1155/2012/939751] [PMID: 22454632]

[19]   Gough DR, Cotter TG. Hydrogen peroxide: a Jekyll and Hyde signalling molecule. Cell Death Dis 2011; 2(10)e213
[http://dx.doi.org/10.1038/cddis.2011.96] [PMID: 21975295]

[20]   Dunnill C, Patton T, Brennan J, *et al.* Reactive oxygen species (ROS) and wound healing: the functional role of ROS and emerging ROS-modulating technologies for augmentation of the healing process. Int Wound J 2017; 14(1): 89-96.
[http://dx.doi.org/10.1111/iwj.12557] [PMID: 26688157]

[21]   Kurahashi T, Fujii J. Roles of Antioxidative Enzymes in Wound Healing. J Dev Biol 2015; 3: 57-70.
[http://dx.doi.org/10.3390/jdb3020057]

[22]   Ranieri SC, Fusco S, Panieri E, *et al.* Mammalian life-span determinant p66shcA mediates obesity-induced insulin resistance. Proc Natl Acad Sci USA 2010; 107(30): 13420-5.
[http://dx.doi.org/10.1073/pnas.1008647107] [PMID: 20624962]

[23]   Berry A, Cirulli F. The p66(Shc) gene paves the way for healthspan: evolutionary and mechanistic

perspectives. Neurosci Biobehav Rev 2013; 37(5): 790-802.
[http://dx.doi.org/10.1016/j.neubiorev.2013.03.005] [PMID: 23524280]

[24]   Paneni F, Mocharla P, Akhmedov A, *et al.* Gene silencing of the mitochondrial adaptor p66(Shc) suppresses vascular hyperglycemic memory in diabetes. Circ Res 2012; 111(3): 278-89.
[http://dx.doi.org/10.1161/CIRCRESAHA.112.266593] [PMID: 22693349]

[25]   Nunes SF, Figueiredo IV, Pereira JS, *et al.* Monoamine oxidase and semicarbazide-sensitive amine oxidase kinetic analysis in mesenteric arteries of patients with type 2 diabetes. Physiol Res 2011; 60(2): 309-15.
[PMID: 21114364]

[26]   Kaludercic N, Giorgio V. The Dual Function of Reactive Oxygen/Nitrogen Species in Bioenergetics and Cell Death: The Role of ATP Synthase. Oxid Med Cell Longev 2016; 20163869610
[http://dx.doi.org/10.1155/2016/3869610] [PMID: 27034734]

[27]   Xia Z, Vanhoutte PM. Nitric oxide and protection against cardiac ischemia. Curr Pharm Des 2011; 17(18): 1774-82.
[http://dx.doi.org/10.2174/138161211796391047] [PMID: 21631419]

[28]   Félétou M, Köhler R, Vanhoutte PM. Nitric oxide: orchestrator of endothelium-dependent responses. Ann Med 2012; 44(7): 694-716.
[http://dx.doi.org/10.3109/07853890.2011.585658] [PMID: 21895549]

[29]   General Considerations.Studies on Respiratory Disorders, Oxidative Stress in Applied Basic Research and Clinical Practice, Springer Science+Business Media New York. 2014; pp. 27-47.

[30]   Tarnow L, Hovind P, Teerlink T, Stehouwer CD, Parving HH. Elevated plasma asymmetric dimethylarginine as a marker of cardiovascular morbidity in early diabetic nephropathy in type 1 diabetes. Diabetes Care 2004; 27(3): 765-9.
[http://dx.doi.org/10.2337/diacare.27.3.765] [PMID: 14988299]

[31]   Lee JH, Park GH, Lee YK, Park JH. Changes in the arginine methylation of organ proteins during the development of diabetes mellitus. Diabetes Res Clin Pract 2011; 94(1): 111-8.
[http://dx.doi.org/10.1016/j.diabres.2011.07.005] [PMID: 21855157]

[32]   Abhary S, Kasmeridis N, Burdon KP, *et al.* Diabetic retinopathy is associated with elevated serum asymmetric and symmetric dimethylarginines. Diabetes Care 2009; 32(11): 2084-6.
[http://dx.doi.org/10.2337/dc09-0816] [PMID: 19675197]

[33]   Korandji C, Zeller M, Guilland JC, *et al.* Time course of asymmetric dimethylarginine (ADMA) and oxidative stress in fructose-hypertensive rats: a model related to metabolic syndrome. Atherosclerosis 2011; 214(2): 310-5.
[http://dx.doi.org/10.1016/j.atherosclerosis.2010.11.014] [PMID: 21146169]

[34]   Van Laer K, Hamilton CJ, Messens J. Low-molecular-weight thiols in thiol-disulfide exchange. Antioxid Redox Signal 2013; 18(13): 1642-53.
[http://dx.doi.org/10.1089/ars.2012.4964] [PMID: 23075082]

[35]   Baez NO, Reisz JA, Furdui CM. Mass spectrometry in studies of protein thiol chemistry and signaling: opportunities and caveats. Free Radic Biol Med 2015; 80: 191-211.
[http://dx.doi.org/10.1016/j.freeradbiomed.2014.09.016] [PMID: 25261734]

[36]   Nagy P. Kinetics and mechanisms of thiol-disulfide exchange covering direct substitution and thiol oxidation-mediated pathways. Antioxid Redox Signal 2013; 18(13): 1623-41.
[http://dx.doi.org/10.1089/ars.2012.4973] [PMID: 23075118]

[37]   Paulsen CE, Carroll KS. Cysteine-mediated redox signaling: chemistry, biology, and tools for discovery. Chem Rev 2013; 113(7): 4633-79.
[http://dx.doi.org/10.1021/cr300163e] [PMID: 23514336]

[38]   Klomsiri C, Karplus PA, Poole LB. Cysteine-based redox switches in enzymes. Antioxid Redox Signal 2011; 14(6): 1065-77.

[http://dx.doi.org/10.1089/ars.2010.3376] [PMID: 20799881]

[39]   Veskoukis AS, Tsatsakis AM, Kouretas D. Dietary oxidative stress and antioxidant defense with an emphasis on plant extract administration. Cell Stress Chaperones 2012; 17(1): 11-21.
[http://dx.doi.org/10.1007/s12192-011-0293-3] [PMID: 21956695]

[40]   Bouayed J, Bohn T. Exogenous antioxidants--Double-edged swords in cellular redox state: Health beneficial effects at physiologic doses *versus* deleterious effects at high doses. Oxid Med Cell Longev 2010; 3(4): 228-37.
[http://dx.doi.org/10.4161/oxim.3.4.12858] [PMID: 20972369]

[41]   Jochmann N, Baumann G, Stangl V. Green tea and cardiovascular disease: from molecular targets towards human health. Curr Opin Clin Nutr Metab Care 2008; 11(6): 758-65.
[http://dx.doi.org/10.1097/MCO.0b013e328314b68b] [PMID: 18827581]

[42]   Xu DP, Li Y, Meng X, *et al.* Natural antioxidants in foods and medicinal plants: extraction, assessment and resources. Int J Mol Sci 2017; 18(1): 80-96.
[http://dx.doi.org/10.3390/ijms18010096] [PMID: 28067795]

[43]   Sankar P, Subhashree S, Sudharani S. Effect of Trigonella foenum-graecum seed powder on the antioxidant levels of high fat diet and low dose streptozotocin induced type II diabetic rats. Eur Rev Med Pharmacol Sci 2012; 16(3) (Suppl. 3): 10-7.
[PMID: 22957413]

[44]   Szkudlinska MA, von Frankenberg AD, Utzschneider KM. The antioxidant N-Acetylcysteine does not improve glucose tolerance or β-cell function in type 2 diabetes. J Diabetes Complications 2016; 30(4): 618-22.
[http://dx.doi.org/10.1016/j.jdiacomp.2016.02.003] [PMID: 26922582]

[45]   Ighodaro OM, Akinloye OA. First line defense antioxidants-superoxide dismutase (SOD), catalase (CAT) and glutathione peroxidase (GPx): Their fundamental role in the entire antioxidant defense grid. Alex J Med 2017; p. 13.

[46]   Gill SS, Tuteja N. Reactive oxygen species and antioxidant machinery in abiotic stress tolerance in crop plants. Plant Physiol Biochem 2010; 48(12): 909-30.
[http://dx.doi.org/10.1016/j.plaphy.2010.08.016] [PMID: 20870416]

[47]   Karuppanapandian T, Moon JC, Kim C, Manoharan K, Kim W. Reactive oxygen species in plants: their generation, signal transduction, and scavenging mechanisms. Aust J Crop Sci 2011; 5: 709.

[48]   Lebovitz RM, Zhang H, Vogel H, *et al.* Neurodegeneration, myocardial injury, and perinatal death in mitochondrial superoxide dismutase-deficient mice. Proc Natl Acad Sci USA 1996; 93(18): 9782-7.
[http://dx.doi.org/10.1073/pnas.93.18.9782] [PMID: 8790408]

[49]   Dayal S, Baumbach GL, Arning E, Bottiglieri T, Faraci FM, Lentz SR. Deficiency of superoxide dismutase promotes cerebral vascular hypertrophy and vascular dysfunction in hyperhomocysteinemia. PLoS One 2017; 12(4)e0175732
[http://dx.doi.org/10.1371/journal.pone.0175732] [PMID: 28414812]

[50]   Surai PF. Selenium in nutrition and health. Nottingham: Nottingham University Press 2006.

[51]   Birben E, Sahiner UM, Sackesen C, Erzurum S, Kalayci O. Oxidative stress and antioxidant defense. World Allergy Organ J 2012; 5(1): 9-19.
[http://dx.doi.org/10.1097/WOX.0b013e3182439613] [PMID: 23268465]

[52]   Espinosa-Diez C, Miguel V, Mennerich D, *et al.* Antioxidant responses and cellular adjustments to oxidative stress. Redox Biol 2015; 6: 183-97.
[http://dx.doi.org/10.1016/j.redox.2015.07.008] [PMID: 26233704]

[53]   Lennicke C, Rahn J, Lichtenfels R, Wessjohann LA, Seliger B. Hydrogen peroxide - production, fate and role in redox signaling of tumor cells. Cell Commun Signal 2015; 13: 39-52.
[http://dx.doi.org/10.1186/s12964-015-0118-6] [PMID: 26369938]

[54] Góth L, Nagy T. Acatalasemia and diabetes mellitus. Arch Biochem Biophys 2012; 525(2): 195-200.
[http://dx.doi.org/10.1016/j.abb.2012.02.005] [PMID: 22365890]

[55] Burk RF, Olson GE, Winfrey VP, Hill KE, Yin D. Glutathione peroxidase-3 produced by the kidney binds to a population of basement membranes in the gastrointestinal tract and in other tissues. Am J Physiol Gastrointest Liver Physiol 2011; 301(1): G32-8.
[http://dx.doi.org/10.1152/ajpgi.00064.2011] [PMID: 21493731]

[56] Labunskyy VM, Hatfield DL, Gladyshev VN. Selenoproteins: molecular pathways and physiological roles. Physiol Rev 2014; 94(3): 739-77.
[http://dx.doi.org/10.1152/physrev.00039.2013] [PMID: 24987004]

[57] Noblanc A, Kocer A, Chabory E, et al. Glutathione peroxidases at work on epididymal spermatozoa: an example of the dual effect of reactive oxygen species on mammalian male fertilizing ability. J Androl 2011; 32(6): 641-50.
[http://dx.doi.org/10.2164/jandrol.110.012823] [PMID: 21441427]

[58] Chabory E, Damon C, Lenoir A, et al. Epididymis seleno-independent glutathione peroxidase 5 maintains sperm DNA integrity in mice. J Clin Invest 2009; 119(7): 2074-85.
[http://dx.doi.org/10.1172/JCI38940] [PMID: 19546506]

[59] Forgione MA, Weiss N, Heydrick S, et al. Cellular glutathione peroxidase deficiency and endothelial dysfunction. Am J Physiol Heart Circ Physiol 2002; 282(4): H1255-61.
[http://dx.doi.org/10.1152/ajpheart.00598.2001] [PMID: 11893559]

[60] Maritim AC, Sanders RA, Watkins JB III. Effects of α-lipoic acid on biomarkers of oxidative stress in streptozotocin-induced diabetic rats. J Nutr Biochem 2003; 14(5): 288-94.
[http://dx.doi.org/10.1016/S0955-2863(03)00036-6] [PMID: 12832033]

[61] Matough FA, Budin SB, Hamid ZA, Alwahaibi N, Mohamed J. The role of oxidative stress and antioxidants in diabetic complications. Sultan Qaboos Univ Med J 2012; 12(1): 5-18.
[http://dx.doi.org/10.12816/0003082] [PMID: 22375253]

[62] Santosh N, David H, Chaya N. Role of ascorbic acid in diabetes mellitus: A comprehensive review. J Med Rad Pathol Surgery 2017; 4: 1-3.
[http://dx.doi.org/10.15713/ins.jmrps.79]

[63] Desai CK, Huang J, Lokhandwala A, Fernandez A, Riaz IB, Alpert JS. The role of vitamin supplementation in the prevention of cardiovascular disease events. Clin Cardiol 2014; 37(9): 576-81.
[http://dx.doi.org/10.1002/clc.22299] [PMID: 24863141]

[64] Kma L. Plant extracts and plant-derived compounds: promising players in a countermeasure strategy against radiological exposure. Asian Pac J Cancer Prev 2014; 15(6): 2405-25.
[http://dx.doi.org/10.7314/APJCP.2014.15.6.2405] [PMID: 24761841]

[65] Szymanska R, Pospisil P, Kruk J. Plant derived antioxidants in disease prevention. Oxi Medi Cel Long. 2016.
[http://dx.doi.org/10.1155/2016/1920208]

[66] Rowe L. DNA damage-induced reactive oxygen species: A genotoxic stress response, PhD Thesis, Emory University, Georgia, USA 2009.

[67] Tilethe S, Chourasiy PK, Dhakad RS, Kumar D. Potential of Rutin and Vildagliptin Combination against Alloxan Induced Diabetic Nephropathy in Mice. Res J Pharmaceutical Sci 2013; 2(9): 1-7.

[68] Lee MT, Lin WC, Yu B, Lee TT. Antioxidant capacity of phytochemicals and their potential effects on oxidative status in animals - A review. Asian-Australas J Anim Sci 2017; 30(3): 299-308.
[http://dx.doi.org/10.5713/ajas.16.0438] [PMID: 27660026]

[69] Tsao R, Deng Z. Separation procedures for naturally occurring antioxidant phytochemicals. J Chromatogr B Analyt Technol Biomed Life Sci 2004; 812(1-2): 85-99.
[http://dx.doi.org/10.1016/S1570-0232(04)00764-0] [PMID: 15556490]

[70]   Giacco F, Brownlee M. Oxidative stress and diabetic complications. Circ Res 2010; 107(9): 1058-70.
       [http://dx.doi.org/10.1161/CIRCRESAHA.110.223545] [PMID: 21030723]

[71]   Tangvarasittichai S. Oxidative stress, insulin resistance, dyslipidemia and type 2 diabetes mellitus.
       World J Diabetes 2015; 6(3): 456-80.
       [http://dx.doi.org/10.4239/wjd.v6.i3.456] [PMID: 25897356]

[72]   Baradaran A, Nasri H, Nematbakhsh M, Rafieian-Kopaei M. Antioxidant activity and preventive
       effect of aqueous leaf extract of *Aloe Vera* on gentamicin-induced nephrotoxicity in male Wistar rats.
       Clin Ter 2014; 165(1): 7-11.
       [PMID: 24589943]

[73]   Asmat U, Abad K, Ismail K. Diabetes mellitus and oxidative stress-A concise review. Saudi Pharm J
       2016; 24(5): 547-53.
       [http://dx.doi.org/10.1016/j.jsps.2015.03.013] [PMID: 27752226]

[74]   Cernea S, Dobreanu M. Diabetes and beta cell function: from mechanisms to evaluation and clinical
       implications. Biochem Med (Zagreb) 2013; 23(3): 266-80.
       [http://dx.doi.org/10.11613/BM.2013.033] [PMID: 24266296]

[75]   Catalano KJ, Maddux BA, Szary J, Youngren JF, Goldfine ID, Schaufele F. Insulin resistance induced
       by hyperinsulinemia coincides with a persistent alteration at the insulin receptor tyrosine kinase
       domain. PLoS One 2014; 9(9)e108693
       [http://dx.doi.org/10.1371/journal.pone.0108693] [PMID: 25259572]

[76]   Tang WH, Martin KA, Hwa J. Aldose reductase, oxidative stress, and diabetic mellitus. Front
       Pharmacol 2012; 3: 87.
       [http://dx.doi.org/10.3389/fphar.2012.00087] [PMID: 22582044]

[77]   Förstermann U, Li H. Therapeutic effect of enhancing endothelial nitric oxide synthase (eNOS)
       expression and preventing eNOS uncoupling. Br J Pharmacol 2011; 164(2): 213-23.
       [http://dx.doi.org/10.1111/j.1476-5381.2010.01196.x] [PMID: 21198553]

[78]   Sena CM, Pereira AM, Seiça R. Endothelial dysfunction - a major mediator of diabetic vascular
       disease. Biochim Biophys Acta 2013; 1832(12): 2216-31.
       [http://dx.doi.org/10.1016/j.bbadis.2013.08.006] [PMID: 23994612]

[79]   Gaschler MM, Stockwell BR. Lipid peroxidation in cell death. Biochem Biophys Res Commun 2017;
       482(3): 419-25.
       [http://dx.doi.org/10.1016/j.bbrc.2016.10.086] [PMID: 28212725]

[80]   Bhutia Y, Ghosh A, Sherpa ML, Pal R, Mohanta PK. Serum malondialdehyde level: Surrogate stress
       marker in the Sikkimese diabetics. J Nat Sci Biol Med 2011; 2(1): 107-12.
       [http://dx.doi.org/10.4103/0976-9668.82309] [PMID: 22470243]

[81]   Florkowski C. HbA1c as a diagnostic test for diabetes mellitus – reviewing the evidence. Clin
       Biochem Rev 2013; 34(2): 75-83.
       [PMID: 24151343]

[82]   Li Z, Malla S, Shin B, Li JM. Battle against RNA oxidation: molecular mechanisms for reducing
       oxidized RNA to protect cells. Wiley Interdiscip Rev RNA 2014; 5(3): 335-46.
       [http://dx.doi.org/10.1002/wrna.1214] [PMID: 24375979]

[83]   Aylward GW. Progressive changes in diabetics and their management. Eye (Lond) 2005; 19(10):
       1115-8.
       [http://dx.doi.org/10.1038/sj.eye.6701969] [PMID: 16304592]

[84]   Lee R, Wong TY, Sabanayagam C. Epidemiology of diabetic retinopathy, diabetic macular edema and
       related vision loss. Eye Vis (Lond) 2015; 2: 17.
       [http://dx.doi.org/10.1186/s40662-015-0026-2] [PMID: 26605370]

[85]   Dal S, Sigrist S. The protective effect of antioxidants consumption on diabetes and vascular

complications. Diseases 2016; 4(3): 1-24.
[http://dx.doi.org/10.3390/diseases4030024] [PMID: 28933404]

[86] Nita M, Grzybowski A. The role of the reactive oxygen species and oxidative stress in the pathomechanism of the age-related ocular diseases and other pathologies of the anterior and posterior eye segments in adults. Oxid Med Cell Longev 2016; 20163164734
[http://dx.doi.org/10.1155/2016/3164734] [PMID: 26881021]

[87] Guzman DC, Olguín HJ, García EH, Peraza AV, de la Cruz DZ, Soto MP. Mechanisms involved in the development of diabetic retinopathy induced by oxidative stress. Redox Rep 2017; 22(1): 10-6.
[http://dx.doi.org/10.1080/13510002.2016.1205303] [PMID: 27420399]

[88] Deanfield JE, Halcox JP, Rabelink TJ. Endothelial function and dysfunction: testing and clinical relevance. Circulation 2007; 115(10): 1285-95.
[http://dx.doi.org/10.1161/CIRCULATIONAHA.106.652859] [PMID: 17353456]

[89] Rajendran P, Rengarajan T, Thangavel J, *et al.* The vascular endothelium and human diseases. Int J Biol Sci 2013; 9(10): 1057-69.
[http://dx.doi.org/10.7150/ijbs.7502] [PMID: 24250251]

[90] Paneni F, Beckman JA, Creager MA, Cosentino F. Diabetes and vascular disease: pathophysiology, clinical consequences, and medical therapy: part I. Eur Heart J 2013; 34(31): 2436-43.
[http://dx.doi.org/10.1093/eurheartj/eht149] [PMID: 23641007]

[91] Kalani M. The importance of endothelin-1 for microvascular dysfunction in diabetes. Vasc Health Risk Manag 2008; 4(5): 1061-8.
[http://dx.doi.org/10.2147/VHRM.S3920] [PMID: 19183753]

[92] Tabit CE, Chung WB, Hamburg NM, Vita JA. Endothelial dysfunction in diabetes mellitus: molecular mechanisms and clinical implications. Rev Endocr Metab Disord 2010; 11(1): 61-74.
[http://dx.doi.org/10.1007/s11154-010-9134-4] [PMID: 20186491]

[93] Giugliano D, Ceriello A, Esposito K. Glucose metabolism and hyperglycemia. Am J Clin Nutr 2008; 87(1): 217S-22S.
[http://dx.doi.org/10.1093/ajcn/87.1.217S] [PMID: 18175761]

[94] Aronoff S, Berkowitz K, Shreiner B, Want L. Glucose metabolism and regulation: beyond insulin and glucagon. Diabetes Spectr 2004; 17(3): 183-90.
[http://dx.doi.org/10.2337/diaspect.17.3.183]

[95] Newsholme P, Cruzat VF, Keane KN, Carlessi R, de Bittencourt PI Jr. Molecular mechanisms of ROS production and oxidative stress in diabetes. Biochem J 2016; 473(24): 4527-50.
[http://dx.doi.org/10.1042/BCJ20160503C] [PMID: 27941030]

[96] Guzik TJ, Mussa S, Gastaldi D, *et al.* Mechanisms of increased vascular superoxide production in human diabetes mellitus: role of NAD(P)H oxidase and endothelial nitric oxide synthase. Circulation 2002; 105(14): 1656-62.
[http://dx.doi.org/10.1161/01.CIR.0000012748.58444.08] [PMID: 11940543]

[97] Lobo V, Patil A, Phatak A, Chandra N. Free radicals, antioxidants and functional foods: Impact on human health. Pharmacogn Rev 2010; 4(8): 118-26.
[http://dx.doi.org/10.4103/0973-7847.70902] [PMID: 22228951]

[98] Bisbal C, Lambert K, Avignon A. Antioxidants and glucose metabolism disorders. Curr Opin Clin Nutr Metab Care 2010; 13(4): 439-46.
[http://dx.doi.org/10.1097/MCO.0b013e32833a5559] [PMID: 20495454]

[99] Turdi S, Li Q, Lopez FL, Ren J. Catalase alleviates cardiomyocyte dysfunction in diabetes: role of Akt, Forkhead transcriptional factor and silent information regulator 2. Life Sci 2007; 81(11): 895-905.
[http://dx.doi.org/10.1016/j.lfs.2007.07.029] [PMID: 17765928]

[100] Sarangarajan R, Meera S, Rukkumani R, Sankar P, Anuradha G. Antioxidants: Friend or foe? Asian

Pac J Trop Med 2017; 10(12): 1111-6.
[http://dx.doi.org/10.1016/j.apjtm.2017.10.017] [PMID: 29268965]

[101] Song Y, Cook NR, Albert CM, Van Denburgh M, Manson JE. Effects of vitamins C and E and beta-carotene on the risk of type 2 diabetes in women at high risk of cardiovascular disease: a randomized controlled trial. Am J Clin Nutr 2009; 90(2): 429-37.
[http://dx.doi.org/10.3945/ajcn.2009.27491] [PMID: 19491386]

[102] Tovmasyan A, Weitner T, Sheng H, *et al.* Differential coordination demands in Fe *versus* Mn water-soluble cationic metalloporphyrins translate into remarkably different aqueous redox chemistry and biology. Inorg Chem 2013; 52(10): 5677-91.
[http://dx.doi.org/10.1021/ic3012519] [PMID: 23646875]

[103] Batinic-Haberle I, Tovmasyan A, Spasojevic I. Mn Porphyrin-Based Redox-Active Drugs: Differential Effects as Cancer Therapeutics and Protectors of Normal Tissue Against Oxidative Injury. Antioxid Redox Signal 2018; 29(16): 1691-724.
[http://dx.doi.org/10.1089/ars.2017.7453] [PMID: 29926755]

[104] Batinic-Haberle I, Tome ME. Thiol regulation by Mn porphyrins, commonly known as SOD mimics. Redox Biol 2019.101139
[http://dx.doi.org/10.1016/j.redox.2019.101139] [PMID: 31126869]

[105] Miriyala S, Spasojevic I, Tovmasyan A, *et al.* Manganese superoxide dismutase, MnSOD and its mimics. Biochim Biophys Acta 2012; 1822(5): 794-814.
[http://dx.doi.org/10.1016/j.bbadis.2011.12.002] [PMID: 22198225]

CHAPTER 4

Administration of Nano Drugs in the Treatment of Diabetes Mellitus

Radhika Tippani[1], Rama Narsimha Reddy Anreddy[2,*] and Mahendar Porika[1]

[1] *Department of Biotechnology, Kakatiya University, Warangal, 506009, Telangana State, India*

[2] *Department of Pharmacology, Jyothishmathi Institute of Pharmaceutical Sciences, Ramakrishna Colony, Thimmapur, Karimnagar 505481, Telangana State, India*

Abstract: Diabetes mellitus (DM) is a metabolic disorder which is the most alarming disease of the modern era, which occurs as a result of lack of insulin secretion or reduced insulin secretion or peripheral insulin resistance. Owing to the lifestyle changes, food habitus and stress, it has now become a pandemic. The incidence of diabetes is rapidly increasing worldwide at a dangerous rate. Over the past 30 years, the status of diabetes has changed from being considered as a mild disorder of the elderly to one of the major causes of morbidity and mortality, affecting the youth and middle-aged people. As per the WHO, 171 million cases were reported in 2000 and are expected to increase to 366 million by 2030. DM incidences continuous to rise and pose a serious threat to human health. DM prevalence is increasing due to lifestyle, ethnicity, and age. Insulin has remained the main treatment for Type 1 diabetes and many Type 2 diabetic patients since its discovery, through parenteral insulin administration. Nanoparticles (NPs), which are minute structures ranging from size 1 to 100nm, are being studied for the treatment of various diseases. Considering the versatility of NPs, it also gives hope for better treatment options in diabetes. Different strategies have been used to manipulate insulin by using NPs, such as encapsulated delivery, *etc*. The objective of this chapter is to resolve the issues concerned with the oral delivery of insulin and also to discuss possible routes for the administration and the use of NPs for the best delivery of insulin. Nanotechnology, as a promising field, has opened new ways for the treatment of DM.

Keywords: Diabetes, Nanoparticles, Nanomedicine, Oral therapy, Oral drug delivery.

INTRODUCTION TO THE ADMINISTRATION OF DRUGS BY ORAL ROUTE

The most common route of drug administration is the oral route. Since the gastro-

---
* **Corresponding author Rama Narsimha Reddy Anreddy:** Jyothishmathi Institute of Pharmaceutical Sciences, Beside LMD Police Station-505481, Ramakrishna Colony, Thimmapur, Karimnagar (Telangana State), India; Tel: +91 99084 57927; E-mail: anreddyram@gmail.com

Atta-ur-Rahman (Ed.)

intestinal tract has a highly absorptive surface, the majority of the drugs are administered orally. However, there are hindrances to this route as antigen inspecting and processing cells are found throughout the gastrointestinal tract (GIT) which cause immunogenic destruction of any compound administered for a longer period. Hence, nanoparticles (NPs) are used since they can be adjusted to increase or decrease bioadhesion to the mucosa and target the particular site [1]. Since mucus layers provide high protection for NP penetration, the residing time of NP is under trial for better results. Decreasing the residing time of NPs leads to failure to penetrate the mucus and trapping and clearance of medicine [2]. Human insulin is a protein composed of amino acids, it is basically a dimer having A and B chains linked by disulfide bonds [3]. When taken orally, this protein is degraded in the GIT before its action and absorption. To overcome this, it is given subcutaneously for better results.

Nanotechnology is the latest developing science with unique applications. Amazed by its properties, scientists have been extensively researching NPs for the development of newer options in the treatment of different diseases. NPs are minute structures with desirable properties having size 1-100 nm in any dimension. Due to their size, when compared to the larger molecules, they are better absorbed and uptaken by intestinal epithelium. As a result of minor modifications in the NP surface and hydrophobicity, the transport across the intestinal cells can be enhanced. The surface properties can be modified by nonspecific changes on the apical cell surface or by grafting a particular ligand targeting the intestinal cells [4,5] The features highlighted in the chapter are regarding the oral administration of the drugs using nanotechnology to bring about a change in anti-diabetic therapy and to help in the improvement of therapeutic efficacy.

DM is a metabolic disorder characterized by an impairment in the metabolism of carbohydrates, proteins and lipids leading to hyperglycemia resulting from the insufficiency of insulin or insulin resistance. DM is one of the fastest growing diseases expected to increase to about 366 million cases by year 2030 as per the predictions given by the WHO [6]. The etiology of diabetes varies due to the impairment in insulin production, reduced response to insulin by the body or insulin insensitivity [7]. Based on the etiology, it is categorized into:

TYPE 1: It is characterized by the absence of insulin production which is immune-mediated or idiopathic resulting from the destruction of β cells of the pancreas [7].

TYPE 2: It is characterized by relative insulin deficiency and insulin resistance resulting from genetic, environmental and behavioral risk factors (stress and lifestyle) [8, 9].

Also, there is a condition known as gestational DM, occurring from the changes in the hormones and body state during pregnancy (which usually resolves following delivery) [10].

CLINICAL FEATURES OF DIABETES

Described as 3P, the symptoms of both type 1 and type 2 are polyphagia (excessive desire to eat) polydipsia (intense thirst) and polyuria (frequent urination and increased urinary frequency at night). In addition, there may be symptoms resulting from hyperglycemia and glycosuria, like fatigue, muscle cramps, impairment of vision, constipation and candidiasis [11].

Type 1 diabetes lasting for a longer period causes complications like micro and macrovascular diseases of heart, arteries and peripheral blood vessels [12 - 14]. Also, there is a higher risk of atherosclerosis in type 2 patients who also have other risk factors like hypertension, obesity and hyperlipidemia. Renal changes occur in longstanding uncontrolled diabetes, eventually leading to end-stage renal disease. There are also retinal and ocular changes in DM, such as early cataract and diabetic retinopathy, causing significant morbidity. Opportunistic infections, commonly of bacterial and fungal origin, are also common in DM.

Pathophysiology

Hyperglycemia induces physiological and behavioral responses in the body (as a result of hyperglycemia, insulin secretion is increased in coordination with the brain).

Type 1 DM

It arises from the autoimmune destruction of insulin-producing cells of the pancreas by CD4+ CD8+ T cells and macrophages. The features include immune-competent cells infiltrating pancreatic islets in the presence of islet cell-specific autoantibodies. Moreover, there is an association of the disease with the genes of class 2 MHC. Also, there are alterations in T cell-mediated immunoregulation and autoimmunity. 85% of the patients showed islet cell antibodies and anti-insulin antibodies in their blood even before receiving insulin therapy. In addition, there is an impairment or inappropriate response in glucagon which is not suppressed by hyperglycemia [15].

Type 2 DM

It is the result of two main pathological defects *i.e.* impaired insulin secretion combined with insulin resistance by peripheral tissues. In these cases, insulin secretion is usually normal but does not compensate for the excessive demands resulting from insulin insensitivity. Hence, there is a stage of hyperinsulinemia with peripheral insulin resistance which can be termed as relative insulin insufficiency. This leads to impaired glucose tolerance. An exception to the etiology of type 2 DM is the maturity-onset diabetes of the young (MODY), resulting from Glucokinase gene mutation on chromosome 7p. MODY is, therefore, described as hyperglycemia diagnosed before 25 years and treated for 5 years without insulin (ICA negative) [16].

Insulin Resistance

It is the resistance to the action of insulin-mediated glucose uptake in the muscles and fat with incomplete suppression of glucose output from the liver and impairment of triglyceride uptake by fat cells [17].

COMPLICATIONS

Complications can be Divided into Acute and Chronic Depending on the Duration of Illness

1. **Acute Complications**
   a. Diabetic ketoacidosis
   b. Hyperosmolar, Hyperglycemic, Nonketotic, Coma
   c. Hypoglycemia

2. **Chronic Complications**
   a. Cardiovascular complications, including increased risk of coronary artery disease (CAD) with angina stroke and cardiac failure.
   b. Nerve damage (neuropathy)
   b. 1 peripheral neuropathy (involving tiny blood vessels which nourish nerves causing numbness, burning and pain)
   b. 2 autonomic neuropathy (leading to vomiting, diarrhea, constipation and erectile dysfunction.)
   c. Nephropathy (kidney damage leading to end-stage renal disease)
   d. Retinopathy (eye damage, cataract glaucoma)
   e. Hearing impairment
   f. Opportunistic infections (skin complications, poor blood supply, poor healing leading to amputations)
   g. Alzheimer's disease (increased risk)

h. Complications in mother with GDM leading to fetal macrosomia, preeclampsia and maternal morbidity

DIAGNOSIS OF DIABETES MELLITUS

Screening is the main step in diagnosing DM in patients exhibiting the following risk factors, obesity, family history and hypertension with DM [17,18]. American diabetes association (ADA) recommends screening and diagnosis of DM in patients having clinical recommendations and evidence by using fasting plasma glucose. WHO recommends OGTT (Oral Glucose Tolerance Test) to be performed in patients at risk [19].

RECOMMENDED INVESTIGATIONS

Random Plasma Blood Sugar Test

It can be done anytime and does not require fasting (more than 200mg/ dl indicates DM)

Fasting Plasma Blood Sugar Test

It requires 8 hours fasting before the test (more than 120 mg/dl indicates DM)

Oral Glucose Tolerance Test

When the RBS or FBS tests are inconclusive then this test is done.

Procedure

Fasting glucose level is determined and then 75 g of oral glucose is given and plasma is tested every 30 min for 2-3hours. A 2-hour glucose level of more than 200mg/ dl confirms DM [19].

INTRODUCTION TO NANOTECHNOLOGY

Nanotechnology is a branch of science which deals with the process that takes place at the molecular level, being nano scale [20]. It is the fastest-growing technology that helps in overcoming the limitations of drug delivery systems. Delivering therapeutic compounds to desirable sites is the biggest challenge in the conventional drug delivery system. Using NPs, the drugs can be delivered in a controlled manner to the particular site, such as targeted tissue drug delivery, minimizing the toxicity and dosage of the particular drug [21]. Modern drug delivery system involves NPs for the following favorable characteristics; high capacity for transporting the drugs, very large active surface for reaction, suitable for small size particles to cross blood levels, ability to accumulate in the target organ/ tissue and considerable lower toxicity [22]. NPs can be organic or inorganic solid particles. The word nano is derived from a Latin word, nanos, meaning dwarf, thus indicating its size.

The most common nanotechnology systems are vesicles, liposomes, nanocapsules and nanospheres, ceramic NPs, dendrimers, polymeric NPs, carbon nanotubes and NPs of biodegradable polymers.

Vesicles

Vesicles are tiny structures consisting of fluid enclosed in a lipid bilayer. These transport the materials in and out of the cell. Vesicular preparations commonly used in drug delivery are block copolymer vesicles, metallic NPs (gold and silver NPs being used) and polymersome metallic NP preparations.

Liposomes

Liposomes are simple concentric bilayer composed of phospholipids that completely enclose the aqueous content. Phospholipids are amphipathic having an affinity for aqueous and polar moieties that have a hydrophobic tail and hydrophilic end.

Liposomes are readily taken by RES (reticulum endothelial system) and enter the cell by endocytosis. Since liposomes are site-specific, they have a few advantages such as they are suitable for both hydrophilic and hydrophobic drugs. Liposomes increase the therapeutic index of the drug. Liposomes suitable for the controlled release can be stabilized by encapsulation. Since they are suitable to administer *via* many routes, they help in reducing the exposure of toxic drugs to sensitive tissues.

Nanocapsules

Nanocapsules are vesicular systems in which the drug moiety is confined to a cavity or space surrounded by tiny polymeric membranes. Hence they make great material for oral and i.v forms in which a large quantity of drugs can be given in higher concentrations.

Nanospheres

Nanospheres are essentially a matrix system that disperses/distributes a solid drug within the polymer throughout the particle.

Ceramic NPs

Ceramic NPs are usually made of compounds like calcium phosphate, silica, alumina titanium, *etc*. These are biocompatible, very small and have high stability, thus pH and temperature-induced denaturation are protected by the drug molecule. A wide variety of drugs become ineffective due to gastric pH, to overcome these, ceramics NPs appear to be promising. They can also be used as carriers for protein-based drugs that are protected from denaturation by ceramic NPs.

Dendrimers

Dendrimer is generally a macromolecule having a highly branched surface, the structure of which provides functional versatility. Dendrimers can be used for targeted and controlled drug release delivery. Since they have a high drug loading capacity, therefore, a large quantity of tissue targeted drugs can be given.

Polymeric NPs

These are usually based on preparations for administering a large quantity of drugs to the target tissue. The polymer is decomposed to lactic acid and glycolic acid during the Krebs cycle, therefore, naturally occurring polymers, such as collagen and cellulose are used. The most studied polymeric NP is PLGA (Poly (lactic-co-glycolic acid). It is approved by the FDA for drug delivery.

PLGA NPs

PLGA is usually incorporated with other compounds to increase its bioavailability, such as eudragit PLGA, antacid incorporated PLGA NP, Chitosan-PLGA NP.

ANTI-DIABETIC MEDICATION

Insulin and oral hypoglycemic agents are the drugs used to treat DM. The implementation of early treatment is necessary to avoid complications associated with chronic hyperglycemia.

As the disease advances, it causes increased morbidity and disability reducing the lifespan of the patient.

Insulin

Lifelong insulin therapy is the mainstay of treatment in patients with type 1 DM (absolute lack of insulin)

There are four basic forms of insulin

1. Rapid-acting
2. Short-acting
3. Intermediate-acting
4. Long-acting

The main route of insulin administration at present is the subcutaneous (SC) route; however, there are other possible routes Fig. (**1**) of administration which are of academic importance [23].

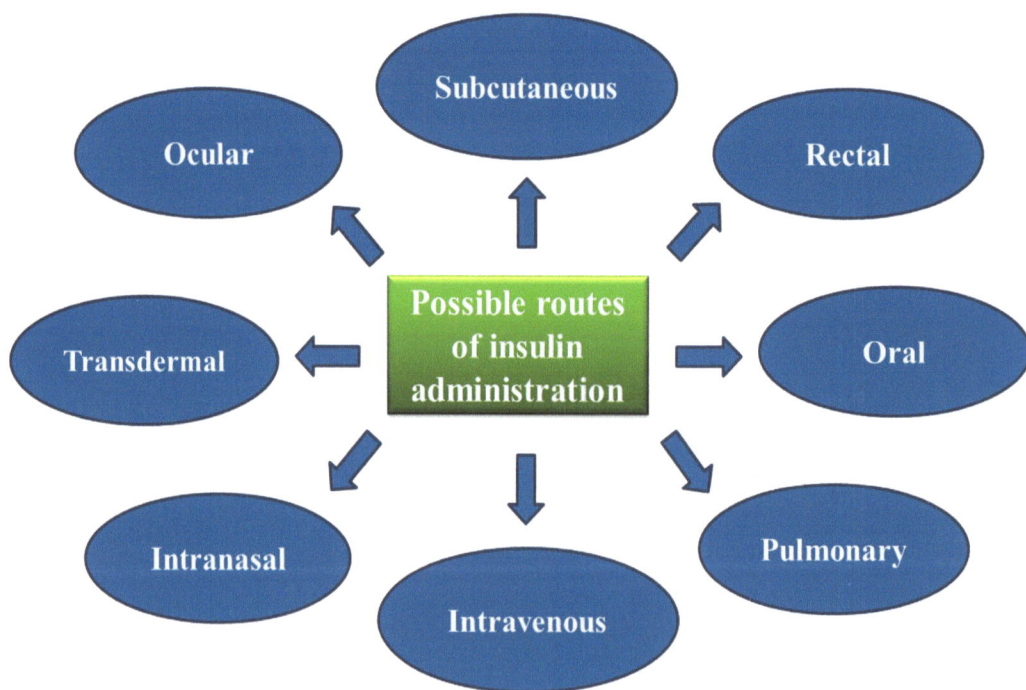

Fig. (1). The key routes of insulin administration.

Considering all the possible routes, oral administration is the most convenient route. It is under research for two decades. Whenever insulin is given through the subcutaneous route, it is directly released into the peripheral circulation and does not follow normal physiology, leading to peripheral hyperinsulinemia and weight gain and thus making it directly proportional to insulin intake. In comparison, when oral insulin is given, it has a physiological advantage since it enters the hepatic portal circulation and acts on the target organ where it inhibits hepatic glucose production [24,25].

Nanocarrier Based Insulin Delivery Systems

As explained above, due to the drawbacks of conventional SC insulin, nanocarriers have been developed with targeting ligands to form targeted insulin-loaded nanocarriers. The major types under study are insulin with chitosan-coated NP, PLGA insulin NP, PACA insulin NP, solid lipid insulin NP, polymeric NP and ceramic NPs. The nanocarrier based drug delivery is meant for i.v., pulmonary route and oral route.

Since the time its oral administration was considered, insulin has been ineffective due to gastric acidity, proteolytic enzymes, as well as chemical and absorptive barriers [26]. To overcome these issues, NPs systems have been studied extensively. Different formulations of NPs like liposomes, microspheres and microemulsion have been used to combat the GIT barrier.

Chitosan is a linear polysaccharide composed of glucosamine. It is coated with NP so as to enhance the permeation of hydrophilic molecules across the intestinal epithelium [27]. Chitosan adheres to the mucosal surface and opens up the junction between the epithelial cells, thus developing a stable and effective chitosan-based insulin NP delivery system [28]. Earlier, the polyelectrolyte complexes of chitosan and insulin easily broke down in the acidic pH of stomach resulting in low pharmacological availability of insulin Fig. (**2**). Now, these NPs are magnetized and are synthesized in a matrix form for a SC or implant-based approach.

PLGA particles in combination with insulin phospholipid complex have been studied. This oral preparation caused prolonged hypoglycemia in diabetic rats and relative bioavailability in comparison to SC insulin was 7.7%. PLGA and Eudragit RS NP are polycationic mucoadhesive acrylic polymers. In this study, hard gelatin capsules were given orally, resulting in prolonged hypoglycemia and around 9.2% bioavailability of insulin. Similarly, antacid incorporated PLGA NP exhibited better oral bioavailability and dose-dependent hypoglycemia in diabetic rats [30 - 32].

Wu *et al.* [33] reported that novel PLGA/HP55 NPs with pH-sensitive characteristics have been prepared for oral insulin delivery using the MESE technique Fig. (**3**). The PLGA/HP55 NPs showed tremendous ability to trap insulin and positive behavior to pH-sensitive release. Additionally, the pharmacodynamic and pharmacokinetic assessment of orally administered PLGA/HP55 NPs in diabetic rats suggested that insulin was rapidly absorbed in the upper intestine and had a substantial hypoglycemic impact. These findings proposed that the PLGA/HP55 NPs, established in the research, could be used as a prospective strategy for multiple daily oral insulin delivery [33]. In addition to all these efforts, few pharmaceutical companies have provided a system to protect the insulin molecule while in the transit period of GIT, *i.e.* encapsulation (physical or NP based) or modification of insulin, making it resistant to degradation. In Ireland, pharmaceutical companies have used the GIPET system which uses matrices of medium-chain fatty acids for enhancing the absorption. The matrices are usually derived from normal dietary components that are safe for longer periods of administration. Therefore, it is a physically mixed formulated tablet designed to be specifically released in the duodenum. India Biocon Ltd has developed a modified form of insulin that possesses physiochemical properties which allow it to stand against enzymatic barriers in the GIT, thus helping in the absorption. IN-105 is recombinant insulin conjugated covalently with a short-chain MPEG derivative. The study is promising in the treatment of diabetes.

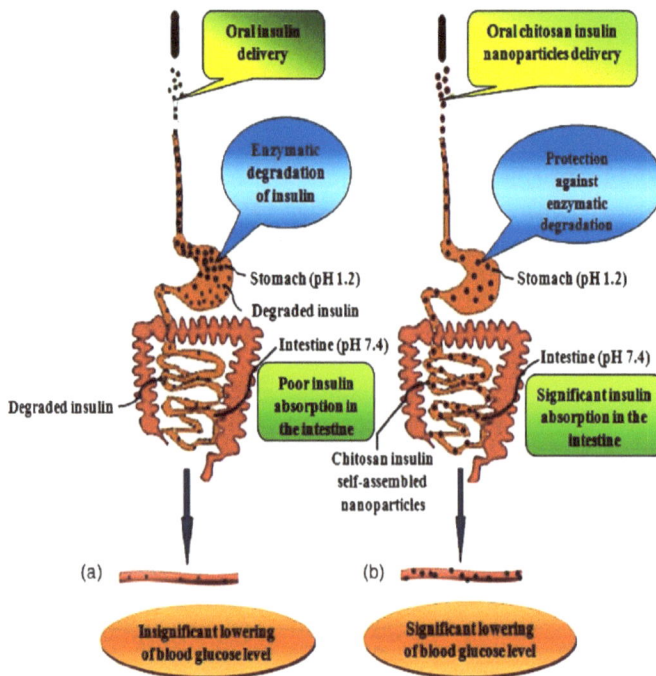

Fig. (2). *In vivo* effectiveness of orally delivered insulin and self-assembled NPs containing chitosan/insulin [29].

Diasome pharmaceuticals have prepared a hepatic directed vesicle for delivering insulin. The HDV is the combination of liposome which is almost 150 nm in size encapsulating the insulin which contains the hepatocyte targeting molecule in the lipid bilayer. The targeting molecule is used to target the delivery of encapsulated insulin to liver cells, hence small amounts of insulin are required to get the desired effect.

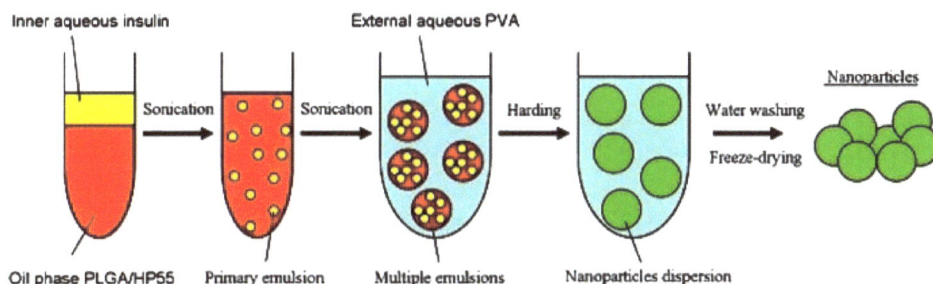

Fig. (3). Insulin loaded PLGA / HP55 NPs preparation [33].

In addition to oral insulin therapy, a new study has emerged on controlling hyperglycemia. This is in view of the reduction in metabolic abnormalities and chorionic complications arising from ROS and reactive nitrogen species [34]. Barath *et al.* [35] used gold NPs as oxidative and anti-hyperglycemic agents in diabetic mice. In the study, it was found that AU NPs controlled several enzymes and reduced ROS generation as well as controlled the lipids, indicating the effectiveness of AU NPs in the improvement of organ functions and their beneficial effects during hyperglycemia. The biological activity of AU NP was attributed to the inhibition of lipid peroxidation. To further prove the activities of AU NPs, Daisy *et al.* [36] also studied them in diabetic rats. A full metabolic study was conducted on the treatment with AU NP. The study revealed the beneficial effects on improving muscle wasting which in turn immensely improved the activity and renal function. It had a beneficial effect on lowering blood glucose and also had anti-inflammatory effects.

Hsieh *et al.* [37] studied the inhibitory effect of AU NP on the fibrillogenesis process of insulin fiber, thus protecting from amyloid degeneration. It has also proved to be promising in the designing of therapies for amyloid-related diseases.

ORAL HYPOGLYCEMIC AGENTS IN THE MANAGEMENT OF TYPE 2 DIABETES MELLITUS

Type 2 diabetes is characterized by the insufficient production of insulin due to insulin resistance and insensitivity. Oral hypoglycemic agents act in order to improve insulin sensitivity in the target tissues. The types of oral hypoglycemic agents are presented in Table **1**.

Nanotechnology is a recent trend in the management of many diseases. Considering its versatility and properties, new drug therapies have been developed which have more advantages over traditional systems. Nanotechnology offers the benefit of controlled, sustained and uniform release of drugs which are more efficacious than their former forms.

Table 1. Classification of oral hypoglycemic agents.

Class	Generic Name	Mechanism of Action	When to Take it	Adverse Effects
Sulfonylureas	Gliclazide Glimepiride Glyburide	Stimulate pancreas to stimulate more insulin	Before meals; Do not take at bedtime	hypoglycemia
Biguanides	Metformine	Reduce the production of glucose by the liver	During meals	Diarrhea, metallic after taste
Thiazolidinediones	Pioglitazones Rosiglitazone	Increase insulin sensitivity & reduce the production of glucose by the liver	With or without food, at the same time each day	Swelling, weight gain
Meglitinides	Nateglinide Repaglinide	Stimulate pancreas to stimulate more insulin	Before meals; Do not take at bedtime	hypoglycemia
Alpha-glucosidases inhibitors	Acarbose	Reduce the absorption of carbohydrates	With meals	Flatulence
DPP-IV inhibitors	Linagliptin Saxagliptin	Intensify the effects of intestinal hormones that regulate carbohydrates	With or without food, at the same time each day	Pharyngitis, headache
GLP-1 agonists	Liraglutide	Intensify the effects of intestinal hormones that regulate carbohydrates	Before meals	Nausea, diarrhea

Oral hypoglycemic agents can be delivered using nanotechnology for efficient treatment of DM.

Various routes of administration that can be used for the administration of oral hypoglycemic agents in combination with NP are

1. Oral
2. Transdermal
3. Implant-based

In the class of sulfonylureas given in Table **1**, glimepiride has been extensively studied for nanotherapy.

Glimepiride is one of the third generation sulfonylurea drugs that stimulates insulin release from β-cells of islets of pancreas and improves the sensitivity of peripheral receptors to insulin. The drug has poor aqueous solubility, slow dissolution rate and poor elimination half-life. To overcome these, NP-based sustained-release drugs are studied using Eudragit RL100 (polymer) and chitosan polymer [38]. Polymeric NPs are preparations consisting of a polymer having NPs distributed in the polymeric matrix. They range from 1 to 50 nm in size in one dimension. They are preferred because they have a higher surface area, which improves the interaction with the surface, has more strength compared to the basic molecule and is more heat resistance. NPs containing an anti-diabetic drug core have been prepared and studied, such as glimepiride. The results show a slow and constant release of the drug from the NPs, thus maintaining a constant drug plasma concentration that improves its therapeutic efficacy. Therefore, the developed formulation overcomes the drawbacks of glimepiride. To improve drug efficacy, a new drug, RL 100, has been assessed for *in vitro* studies and stability studies which have been so far successful and are ready for *in vivo* and preclinical studies. In another study, glimepiride was assessed by Yadav *et al.* [39] for nanosuspensions. Nanosuspensions are colloidal dispersions of nano-sized drug particles stabilized by surface-active agents (surfactant). It consists of a pure drug with a vehicle in which the suspended particles are less than 1 μm in size. Nanosuspensions can also be lyophilized or can be incorporated into a solid matrix. Therefore, for the preparation of glimepiride nanosuspension, an ERLPO polymer and a specific quantity of glimepiride were used. Hence in the *in vivo* studies, nanosuspension showed a significantly higher Cmax, *i.e.* an increase in the saturation solubility of NP as they are absorbed without the initial time-consuming dissolution step. As a result, the plasma value of nanosuspension was found to be significantly higher than the drug suspension, indicating an improvement in the relative bioavailability of the drug in nanosuspension.

In another study [40], Zein NP s /PLGA triblock *in situ* forming implants using glimepiride were studied. As we know, glimepiride is practically insoluble in water, therefore, it falls under VCS class II drugs. When given orally, it has irregular and low bioavailability. In an attempt to overcome these hurdles, transdermal/intramuscular/ subcutaneously route has been considered. Since surfactants cause undesirable skin sensitivity, therefore intramuscular or subcutaneously route is preferred. PGLA-based solid biodegradable microparticles in a depot gel form have been used as implants either injected or incorporated intramuscularly/ subcutaneously using aseptic techniques. The depot releases the drug in a controlled manner, thus continuously acting on the target site leading to a sustained drug release. This gives better glycemic control avoiding complications. In the formulation, optimized Zein-based NPs embedded within the PLGA-PEG-PLGA copolymer (creating tri-copolymer) were used. The gel was injected intramuscularly for studies. Zein is a protein derived from maize which has good biocompatibility and is nontoxic [41]. After the implantation, it allows uniform dispersion of the drug with acceptable patient compliance. In the tri-blocking implant, the drug release can be adjusted by increasing or decreasing the concentration of tri-block copolymers. Thus, in a controlled manner, the drug release rate is adjusted and the gel is implanted, proving to be effective in the treatment of DM in contrast to the burst release of drugs from other routes of administration [40]. The gel is implanted using a 21gage needle making it easy for administration.

Liposomes have been used successfully to transfer drugs across the skin. Singh *et al.* [42] studied the anti-diabetic efficacy of glibenclamide loaded liposomes in diabetic rats. Glibenclamide is a sulfonylurea that acts by binding and inhibiting the ATP sensitive potassium channels inhibitory regulatory subunit sulfonylurea receptor 1 in β cells of the pancreas. Thus, by stimulating insulin release, the anti-hyperglycemic effects of various loaded liposomes of glibenclamide were assessed at different time intervals on fasting blood glucose levels in diabetic rats. The blood glucose reduced at 2nd hour and reached normal at 4th and 8th hour which was maintained up to 16th hour.

Raja *et al.* [43] reported the evaluation and formulation of maltodextrin-based proniosomal drugs consisting of glipizide as an anti-diabetic medication. Niosomes are usually prepared from nontoxic and biodegradable lipids and nonionic surfactants. They can also be used for parenteral ocular and transdermal drug delivery. A glipizide-loaded maltodextrin-based proniosome was prepared and the formulation was assessed for effectiveness of entrapment. It showed greater effectiveness of entrapment and the release of drug by 99.2% at the end of 24 hours. They are being further evaluated in diabetic rats.

Metformin is the most commonly used drug in the treatment of DM. Metformin improves hyperglycemia by suppressing hepatic glucose production (gluconeogenesis). Polymeric NPs in combination with metformin have been studied by various research groups. Such preparation is only significant than the metformin preparation as the bioavailability of metformin is nearly 50 to 60% under fasting conditions [44].

Polymeric NP in combination with PMMA and PLGA polymer has been used. In doing so, the bioavailability of the drug improved along with its efficiency. In addition, toxicology studies also been conducted to rule out the adverse effects of adding NP to the conventional drug. Studies indicated no increase in the toxicity on adding NP to the preparation. In addition, no hepatotoxicity or renal toxicity was found when such a compound was administered. The only result was an increase in the bioavailability of the drug on oral administration without any toxicity [45].

The other nano preparations for metformin are microspheres and microcapsules. They used micro size polymeric particles as the drug delivery system. Microspheres consist of spherical micrometric matrix containing a core drug. Similarly, microcapsules are the systems that encapsulate the core drug. The three preparations studied are metformin HCL loaded microcapsules, Eudragit RL 100 microcapsule and metformin HCL containing ethyl cellulose microspheres [46]. The drug release from the microcapsule was reported to be faster (77%) and the efficacy was 87-89%, thus controlling the plasma glucose levels rapidly and efficiently. The drug release from the microspheres was 72-87%, therefore, the glucose level was controlled within 30min of drug administration and could be maintained up to 12 hours, further reducing the repeated administration of metformin by replacing it by a single effective dose.

Rosiglitazone belongs to the class of thiazolidinedione, an anti-diabetic drug that has a short half-life but high insulin sensitivity. It has major side effects, including hepatic toxicity, anemia, GI disturbances and edema. Gelatin NPs loaded with rosiglitazone were studied for encapsulation and entrapment. This study was conducted by Vandana *et al.* [47]. Rosiglitazone loaded gelatin NP showed entrapment efficiency of 90% varying with drug-polymer ratio. The drug was studied for *in vitro* release kinetics. It showed a biphasic pattern of drug release by initial burst and later, slow diffusion of the drug from the matrix. In practical situations, it applied to reduce the toxicity of lesser frequent dosage and prolonged the duration of action. The drug was completely released in a controlled manner, *i.e.* 80% at the end of 32 hours.

Enkhzaya and Jeong-Sook [48] prepared a nano-emulsion preparation in combination with rosiglitazone. Emulsions with a droplet size on the nanometer scale *i.e.* ranging from 20-200 nm are referred as nano-emulsions. Nano-emulsions are isotropic disperse systems of two non-miscible liquids, usually an oily system dispersed in an aqueous system or an aqueous system dispersed in an oily system forming droplets or other oily phases of nano metric sizes. In this preparation, rosiglitazone is loaded in a cationic lipid emulsion. The result showed the improvement in the *in vitro* drug release in comparison to rosiglitazone alone along with the cellular uptake of rosiglitazone in insulin resistance HepG2 cells. Thus, drug-loaded cationic lipid emulsions are excellent drug delivery systems for rosiglitazone that enhance the cellular uptake efficiency in target cells.

Repaglinide is a fast-acting prandial glucose regulator. It acts by stimulating insulin release from the pancreas. The drug is under evaluation for nano-emulsion technique. The drug is incorporated in the oily phase of nano-emulsion to give improved biopharmaceutical effect. The optimized repaglinide nano-emulsion showed better hypoglycemic effect in comparison to the drug alone in diabetic rats. The drug was found to be stable at different temperatures and humidity on exposure for a period of 3 months [49].

In the treatment of type 2 diabetes mellitus, recently, an incretin-based therapy has been found to be effective worldwide. Glucagon-like peptide released (GLP1) from the L cells of colon and ilium stimulates the insulin release in response to glucose level. Therefore, GLP 1 controlled the blood sugar levels in type 2 diabetes without the risk of hypoglycemia. The clinical usage of GLP 1 is difficult due to its short half-life (2min) due to the fast degradation of the GLP 1by Dipeptidyl peptidase-4 (DPP4). To solve the issue, a DPP4 resistance analog has been developed. Liraglutide is a fatty acid derivative of GLP1 that shares 97% sequence homology with GLP1 [50]. The structural modification was carried out to prolong the half-life and to improve the binding with other molecules. But lira still retains the physiological properties of GLP 1, *i.e.*

1. Stimulation of insulin release.
2. Glucagon suppression in a glucose-dependent manner.
3. Improving β cell function
4. Reducing insulin resistance.
5. Delay in gastric emptying and increasing satiety.

Victoza (the solution injection of lira) developed by novonordisc is FDA and EMI approved in the treatment of type 2 DM. Single subcutaneous injection is needed daily for the patients. To make lira long acting, I has been combined with thermo gelling polymers and has been extensively studied. In this study, long-acting delivery systems of lira in the treatment of type 2 diabetes mellitus using thermo gelling block copolymers were assessed [51].

Among the polymers undergoing a thermo-reversible sol-gel transition in water, block copolymers composed of hydrophobic polyesters, such as poly(lactic acid-co-glycolic acid) (PLGA), poly(ε-caprolactone) (PCL), poly (ε-caprolactone--o-glycolic acid) (PCGA), and hydrophilic PEG are particularly interesting and important because of the good safety profile and facile synthesis [52 - 54].

The *in vitro* release of lira from thermo gel systems was evaluated. The release of lira from the PLGA-PEG-PLGA hydrogel showed a low burst than with 17% of the loaded amount of the drug released on the first day with significant incomplete release observed in the later stages making 56% of the drug to be released within 9 days, therefore, indicating the sustained release profile of the drug over 9 days and cumulative amount of release *i.e.* 85%. As a result, the thermo gel was found to be effective in releasing the drug constantly and significantly, making it suitable for long-acting delivery. In addition to the above study, an *in vivo* hypoglycemic study was also conducted. This study showed similar results after the administration of thermo gel formulation in rats. An oral gavage of glucose was administered and the glucose levels were monitored from day 0 to day 7. These tests indicated that the gel formulations of lira had significant hypoglycemic effects up to one week in mice after single SC administration. In addition, they also showed good biocompatibility without any toxicity of the implanted biomaterial [51].

In another study, liraglutide loaded multivesicular liposomes (MVL) were studied for sustained drug delivery systems by Zhang *et al.* [55]. MVL is a lipid-based sustained-release system with biodegradable and biocompatible properties since it is derived from naturally occurring lipids. MVL particles are composed of nonconcentric multiple lipid layers with particles ranging from several to 10 micrometers. Research shows the arrangement of the lipid layer is responsible for the increased level of stability and for reducing the burst release of the drug and prolonging the release from few days to few weeks. Liraglutide was encapsulated with good efficiency of 82% into MVL particles. The drug release from the MVL was 16% in the first 4 hours and a sustained release of 168 hours of liraglutide loaded MVL without a burst release. The fluorescent assays showed the retention of LRGMVL in the subcutaneous tissue until 168 hours post-SC injection. The sustained delivery of LRG MVLs prolonged the glucose-lowering effect significantly and maintained a therapeutic, steady and constant drug level for a week after a single injection in live rats. This result indicates that RLGMVL can be successfully delivered by liraglutide sustained drug delivery systems.

LIMITATIONS

Even though many attempts have been made to develop an effective NP oral insulin therapy, the formulation synthesis is not cost-effective, creating loopholes in the production of efficient nano-drugs with commercial significance. Oral insulin is under study for nearly two decades but as far as clinical trials are concerned, it is of limited significance and majority have faced failure. The oral bioavailability, nanocarrier biocompatibility and immune response to nano drugs are yet to be studied. A few studies regarding the toxicology of nano-formulations are conducted which are still inconclusive. *In vitro* studies and animal models do not provide the features of mucosal interaction, enzymatic activity and immune reaction. More clinical trials and *in vivo* studies are needed to demonstrate the effects of nanomedicine as proven from *in vitro* studies.

CONCLUSION

• Targeted orally administered NPs have significant prospective biomedical application and several putative benefits for the delivery of oral drugs, such as the protection of fragile drugs or the potential for drug pharmacokinetics alteration. Despite these benefits, the oral delivery of drugs by NPs remains challenging. To achieve efficient drug delivery, NPs must i) avoid rapid mucus clearance; ii) penetrate the mucus layer; and iii) be extensively taken up by the intestinal epithelium. It has been shown that optimizing the particle size and surface characteristics and targeting particular cells through ligand grafting will improve NP transport across the intestinal epithelium. The primary benefit of these particles is that their binding characteristics are enhanced by the use of nondegraded ligands in the GI tract, unlike peptide / proteinic ligands, and are not restricted to receptors that are only associated with proteins.

• Nanotechnology is expected to play a significant role in enhancing complication management, such as diabetes, over the next century. It is encouraging to develop formulations with FDA-approved nanotechnology combined with the clinical achievement of insulin-providing techniques through the oral route.With each passing day, the effect of nanotechnology on medicine is increasing. Although nanomedicine is still in its infancy, efficient diabetes therapy is one of the main prospective applications. Oral insulin, in specific, could be promising, particularly as it appears to have progressed with nanotechnology studies as a treatment, enabling several kinds of encapsulations to bypass the gastric acid environment.

• It is hoped, with progressively advanced models and integrations, to see the growth of personalized diabetic therapy based on NPs in the near future.

• The drug delivery mechanism based on NP currently holds an important position in the pharmaceutical industry. A fresh drug delivery scheme of a current drug can provide an economically significant fresh marketability. Insulin based on next-generation NPs may be the future type 1 DM medicine. This nanocarrier-based insulin supply could replace the trading scenario in the near future

CONSENT FOR PUBLICATION

Not applicable.

CONFLICT OF INTEREST

The authors confirm that the contents of this chapter have no conflict of interest.

ACKNOWLEDGEMENTS

Declare none.

REFERENCES

[1]   Viscido A, Capannolo A, Latella G, Caprilli R, Frieri G. Nanotechnology in the treatment of

inflammatory bowel diseases. J Crohn's Colitis 2014; 8(9): 903-18.
[http://dx.doi.org/10.1016/j.crohns.2014.02.024] [PMID: 24686095]

[2]   Ensign LM, Cone R, Hanes J. Oral drug delivery with polymeric nanoparticles: the gastrointestinal mucus barriers. Adv Drug Deliv Rev 2012; 64(6): 557-70.
[http://dx.doi.org/10.1016/j.addr.2011.12.009] [PMID: 22212900]

[3]   Li J, Rossetti G, Dreyer J, *et al.* Molecular simulation-based structural prediction of protein complexes in mass spectrometry: the human insulin dimer. PLOS Comput Biol 2014; 10(9)e1003838
[http://dx.doi.org/10.1371/journal.pcbi.1003838] [PMID: 25210764]

[4]   Jiang Y, Huo S, Mizuhara T, *et al.* The interplay of size and surface functionality on the cellular uptake of sub-10 nm gold nanoparticles. ACS Nano 2015; 9(10): 9986-93.
[http://dx.doi.org/10.1021/acsnano.5b03521] [PMID: 26435075]

[5]   Salatin S, Yari Khosroushahi A. Overviews on the cellular uptake mechanism of polysaccharide colloidal nanoparticles. J Cell Mol Med 2017; 21(9): 1668-86.
[http://dx.doi.org/10.1111/jcmm.13110] [PMID: 28244656]

[6]   Krol S, Ellis-Behnke R, Marchetti P. Nanomedicine for treatment of diabetes in an aging population: state-of-the-art and future developments. Maturitas 2012; 73(1): 61-7.
[http://dx.doi.org/10.1016/j.maturitas.2011.12.004] [PMID: 22209199]

[7]   Maitra A, Abbas AK. Endocrine system Robbins and Cotran Pathologic basis of disease (7thedtn). Philadelphia: Saunders 2005; pp. 1156-226.

[8]   Wild S, Roglic G, Green A, Sicree R, King H. Global prevalence of diabetes: estimates for the year 2000 and projections for 2030. Diabetes Care 2004; 27(5): 1047-53.
[http://dx.doi.org/10.2337/diacare.27.5.1047] [PMID: 15111519]

[9]   WHO Expert Committee on Definition. Diagnosis and Classification of Diabetes Mellitus and its Complications. Geneva 1999; pp. 1-59.

[10]   Guidance for Industry. Diabetes Mellitus: Developing Drugs and Therapeutic Biologics for Treatment and Prevention, US Department of Health and Human Services Food and Drug Administration Center for Drug Evaluation and Research. CDER 2008; p. 3.

[11]   Bearse MA Jr, Han Y, Schneck ME, Barez S, Jacobsen C, Adams AJ. Local multifocal oscillatory potential abnormalities in diabetes and early diabetic retinopathy. Invest Ophthalmol Vis Sci 2004; 45(9): 3259-65.
[http://dx.doi.org/10.1167/iovs.04-0308] [PMID: 15326149]

[12]   Hove MN, Kristensen JK, Lauritzen T, Bek T. The prevalence of retinopathy in an unselected population of type 2 diabetes patients from Arhus County, Denmark. Acta Ophthalmol Scand 2004; 82(4): 443-8.
[http://dx.doi.org/10.1111/j.1600-0420.2004.00270.x] [PMID: 15291939]

[13]   Saely CH, Aczel S, Marte T, Langer P, Drexel H. Cardiovascular complications in Type 2 diabetes mellitus depend on the coronary angiographic state rather than on the diabetic state. Diabetologia 2004; 47(1): 145-6.
[http://dx.doi.org/10.1007/s00125-003-1274-6] [PMID: 14676943]

[14]   Seki M, Tanaka T, Nawa H, *et al.* Involvement of brain-derived neurotrophic factor in early retinal neuropathy of streptozotocin-induced diabetes in rats: therapeutic potential of brain-derived neurotrophic factor for dopaminergic amacrine cells. Diabetes 2004; 53(9): 2412-9.
[http://dx.doi.org/10.2337/diabetes.53.9.2412] [PMID: 15331553]

[15]   Holt RI. Diagnosis, epidemiology and pathogenesis of diabetes mellitus: an update for psychiatrists. Br J Psychiatry Suppl 2004; 47: S55-63.
[http://dx.doi.org/10.1192/bjp.184.47.s55] [PMID: 15056594]

[16]   Sekikawa A, Tominaga M, Takahashi K, *et al.* Prevalence of diabetes and impaired glucose tolerance in Funagata area, Japan. Diabetes Care 1993; 16(4): 570-4.

[http://dx.doi.org/10.2337/diacare.16.4.570] [PMID: 8462380]

[17]   Cryer PE. Minireview: Glucagon in the pathogenesis of hypoglycemia and hyperglycemia in diabetes. Endocrinology 2012; 153(3): 1039-48.
[http://dx.doi.org/10.1210/en.2011-1499] [PMID: 22166985]

[18]   Harris MI, Klein R, Welborn TA, Knuiman MW. Onset of NIDDM occurs at least 4-7 yr before clinical diagnosis. Diabetes Care 1992; 15(7): 815-9.
[http://dx.doi.org/10.2337/diacare.15.7.815] [PMID: 1516497]

[19]   Gillett MJ. International Expert Committee. International Expert Committee report on the role of the A1C assay in the diagnosis of diabetes. Diabetes Care 2009; 32(7): 1327-34.
[http://dx.doi.org/10.2337/dc09-9033] [PMID: 19502545]

[20]   Bhatia S. Natural Polymer Drug Delivery Systems: Nanoparticles, Plants, and Algae. Springer; 2016.

[21]   Patel A, Khanna S, Xavier GK, Khanna K, Goel B. Polymeric Nano-Particles for Tumor Targeting â A Review. Int J Drug Development and Research 2017; 9: 50-9.

[22]   Bennet D, Kim S. Polymer nanoparticles for smart drug delivery.Application of Nanotechnology in Drug Delivery. InTech 2014.
[http://dx.doi.org/10.5772/58422]

[23]   Golden SH, Sapir T. Methods for insulin delivery and glucose monitoring in diabetes: summary of a comparative effectiveness review. J Manag Care Pharm 2012; 18(6) (Suppl.): S1-S17.
[http://dx.doi.org/10.18553/jmcp.2012.18.s6-A.1] [PMID: 22984955]

[24]   Sonaje K, Lin KJ, Wey SP, *et al.* Biodistribution, pharmacodynamics and pharmacokinetics of insulin analogues in a rat model: Oral delivery using pH-responsive nanoparticles vs. subcutaneous injection. Biomaterials 2010; 31(26): 6849-58.
[http://dx.doi.org/10.1016/j.biomaterials.2010.05.042] [PMID: 20619787]

[25]   Veiseh O, Tang BC, Whitehead KA, Anderson DG, Langer R. Managing diabetes with nanomedicine: challenges and opportunities. Nat Rev Drug Discov 2015; 14(1): 45-57.
[http://dx.doi.org/10.1038/nrd4477] [PMID: 25430866]

[26]   Han L, Zhao Y, Yin L, *et al.* Insulin-loaded pH-sensitive hyaluronic acid nanoparticles enhance transcellular delivery. AAPS PharmSciTech 2012; 13(3): 836-45.
[http://dx.doi.org/10.1208/s12249-012-9807-2] [PMID: 22644708]

[27]   Cheung RC, Ng TB, Wong JH, Chan WY. Chitosan: an update on potential biomedical and pharmaceutical applications. Mar Drugs 2015; 13(8): 5156-86.
[http://dx.doi.org/10.3390/md13085156] [PMID: 26287217]

[28]   Zhang HL, Wu SH, Tao Y, Zang LQ, Su ZQ. Preparation and characterization of water-soluble chitosan nanoparticles as protein delivery system. J Nanomater 2010; 2010: 1.
[http://dx.doi.org/10.1155/2010/651326]

[29]   Mukhopadhyay P, Sarkar K, Chakraborty M, Bhattacharya S, Mishra R, Kundu PP. Oral insulin delivery by self-assembled chitosan nanoparticles: *in vitro* and *in vivo* studies in diabetic animal model. Mater Sci Eng C 2013; 33(1): 376-82.
[http://dx.doi.org/10.1016/j.msec.2012.09.001] [PMID: 25428084]

[30]   Wu ZM, Zhou L, Guo XD, *et al.* HP55-coated capsule containing PLGA/RS nanoparticles for oral delivery of insulin. Int J Pharm 2012; 425(1-2): 1-8.
[http://dx.doi.org/10.1016/j.ijpharm.2011.12.055] [PMID: 22248666]

[31]   Sharma G, van der Walle CF, Ravi Kumar MN. Antacid co-encapsulated polyester nanoparticles for peroral delivery of insulin: development, pharmacokinetics, biodistribution and pharmacodynamics. Int J Pharm 2013; 440(1): 99-110.
[http://dx.doi.org/10.1016/j.ijpharm.2011.12.038] [PMID: 22227604]

[32]   Zhang X, Sun M, Zheng A, Cao D, Bi Y, Sun J. Preparation and characterization of insulin-loaded

bioadhesive PLGA nanoparticles for oral administration. Eur J Pharm Sci 2012; 45(5): 632-8.
[http://dx.doi.org/10.1016/j.ejps.2012.01.002] [PMID: 22248882]

[33]   Wu ZM, Ling L, Zhou LY, *et al.* Novel preparation of PLGA/HP55 nanoparticles for oral insulin delivery. Nanoscale Res Lett 2012; 7(1): 299-05.
[http://dx.doi.org/10.1186/1556-276X-7-299] [PMID: 22682064]

[34]   Si S, Pal A, Mohanta J, Satapathy SS. Gold nanostructure materials in diabetes management. J Phys D Appl Phys 2017; 50(13)134003
[http://dx.doi.org/10.1088/1361-6463/aa5e21]

[35]   Barathmanikanth S, Kalishwaralal K, Sriram M, *et al.* Anti-oxidant effect of gold nanoparticles restrains hyperglycemic conditions in diabetic mice. J Nanobiotechnology 2010; 8(1): 16.
[http://dx.doi.org/10.1186/1477-3155-8-16] [PMID: 20630072]

[36]   Daisy P, Saipriya K. Biochemical analysis of Cassia fistula aqueous extract and phytochemically synthesized gold nanoparticles as hypoglycemic treatment for diabetes mellitus. Int J of nanomed 2012; 7: 1189.
[http://dx.doi.org/10.2147/IJN.S26650]

[37]   Hsieh S, Chang CW, Chou HH. Gold nanoparticles as amyloid-like fibrillogenesis inhibitors. Colloids Surf B Biointerfaces 2013; 112: 525-9.
[http://dx.doi.org/10.1016/j.colsurfb.2013.08.029] [PMID: 24060166]

[38]   Sumalatha N, Umasankar K, Jayachandra Reddy P, Tejaswini P. Formulation and Evaluation of Anti-Diabetes by Nanoparticles. Inter J Pharmacy Review & Research 2015; 5(4): 353-60.

[39]   Yadav SK, Mishra S, Mishra B. Eudragit-based nanosuspension of poorly water-soluble drug: formulation and *in vitro in vivo* evaluation. AAPS PharmSciTech 2012; 13(4): 1031-44.
[http://dx.doi.org/10.1208/s12249-012-9833-0] [PMID: 22893314]

[40]   Ahmed OA, Zidan AS, Khayat M. Mechanistic analysis of Zein nanoparticles/PLGA triblock in situ forming implants for glimepiride. Int J Nanomedicine 2016; 11: 543-55.
[http://dx.doi.org/10.2147/IJN.S99731] [PMID: 26893561]

[41]   Jao D, Xue Y, Medina J, Hu X. Protein-Based Drug-Delivery Materials. Materials (Basel) 2017; 10(5): 517.
[http://dx.doi.org/10.3390/ma10050517] [PMID: 28772877]

[42]   Singh M, Balamurugan M, Gupta A, Yadav S, Sharma A, *et al.* Antidiabetic activity of Glibenclamide loaded liposomes in Alloxan induced diabetic rats. Ars Pharm 2007; 48: 31-6.

[43]   Raja K, Jestin PU, Athul PV, Tamizharasi S, Sivakumar T. Formulation and evaluation of maltodextrin based proniosomal drug delivery system containing anti-diabetic (Glipizide) drug. Int J Pharm Tech Res 2011; 3(1): 471-7.

[44]   Cetin M, Sahin S. Microparticulate and nanoparticulate drug delivery systems for metformin hydrochloride. Drug Deliv 2016; 23(8): 2796-805.
[http://dx.doi.org/10.3109/10717544.2015.1089957] [PMID: 26394019]

[45]   Lekshmi UM, Reddy PN. Preliminary toxicological report of metformin hydrochloride loaded polymeric nanoparticles. Toxicol Int 2012; 19(3): 267-72.
[http://dx.doi.org/10.4103/0971-6580.103667] [PMID: 23293465]

[46]   Nath B, Nath LK, Mazumdar B, *et al.* Design and development of metformin HCl floating microcapsules using two polymers of different permeability characteristics. IJPSN 2009; 2: 627-37.

[47]   Vandana S, Amrendra K. Development and characterization of rosiglitazone loaded gelatin nanoparticles using two step desolvation method. Int J Pharm Sci Rev Res 2010; 5: 100-3.

[48]   Davaa E, Park JS. Formulation parameters influencing the physicochemical characteristics of rosiglitazone-loaded cationic lipid emulsion. Arch Pharm Res 2012; 35(7): 1205-13.
[http://dx.doi.org/10.1007/s12272-012-0711-9] [PMID: 22864743]

[49]   Akhtar J, Siddiqui HH, Fareed S, *et al.* Nanoemulsion: for improved oral delivery of repaglinide Drug delivery 2016; 23(6): 2026-34.

[50]   Prasad-Reddy L, Isaacs D. A clinical review of GLP-1 receptor agonists: efficacy and safety in diabetes and beyond. Drugs Context 2015; 4212283
[http://dx.doi.org/10.7573/dic.212283] [PMID: 26213556]

[51]   Chen Y, Li Y, Shen W, *et al.* Controlled release of liraglutide using thermogelling polymers in treatment of diabetes. Sci Rep 2016; 6: 31593.
[http://dx.doi.org/10.1038/srep31593] [PMID: 27531588]

[52]   Zentner GM, Rathi R, Shih C, *et al.* Biodegradable block copolymers for delivery of proteins and water-insoluble drugs. J Control Release 2001; 72(1-3): 203-15.
[http://dx.doi.org/10.1016/S0168-3659(01)00276-0] [PMID: 11389999]

[53]   Yu L, Sheng WJ, Yang DC, Ding JD. Design of molecular parameters to achieve block copolymers with a powder form at dry state and a temperature-induced sol-gel transition in water without unexpected gelling prior to heating. Macromol Res 2013; 21: 207-15.
[http://dx.doi.org/10.1007/s13233-013-1021-x]

[54]   Yu L, Zhang Z, Zhang H, Ding J. Biodegradability and biocompatibility of thermoreversible hydrogels formed from mixing a sol and a precipitate of block copolymers in water. Biomacromolecules 2010; 11(8): 2169-78.
[http://dx.doi.org/10.1021/bm100549q] [PMID: 20690723]

[55]   Zhang L, Ding L, Tang C, Li Y, Yang L. Liraglutide-loaded multivesicular liposome as a sustained-delivery reduces blood glucose in SD rats with diabetes. Drug Deliv 2016; 23(9): 3358-63.
[http://dx.doi.org/10.1080/10717544.2016.1180723] [PMID: 27099000]

CHAPTER 5

SGLT-2 Inhibitors: An Evidence-Based Perspective

Siddhartha Dutta, Pramod Kumar Sharma* and **Arup Kumar Misra**

All India Institute of Medical Sciences, Jodhpur, Rajasthan, India

Abstract: Diabetes mellitus (DM) is one of the most prevalent diseases of modern society. There are several therapeutic options available, but they also have many shortcomings. With the limitations and pitfalls of the existing therapies of diabetes, there is always a need for better drugs. This review is an attempt to give comprehensive details about the merits and demerits of a class of drugs called SGLT-2 inhibitors. SGLT-2 inhibitors act by increasing glucose excretion through urine and do not have any effect on insulin secretion, therefore, the risk of hypoglycemia is less. SGLT-2 inhibitors that are in clinical use are: dapagliflozin, empagliflozin, canagliflozin, and ertugliflozin. Considering the benefits offered by SGLT-2 inhibitors over existing antidiabetics, they deserve an important place in the therapy of T2DM and are found to be useful in T1DM, as studies have suggested previously. Beneficial effects of these drugs extend beyond controlling hyperglycemia, *e.g.*, reduction in body weight, reduction in blood pressure and a proven and appreciable reduction in cardiovascular adverse events, maintenance of arterial elasticity and decrease in visceral adipose tissue deposition. The demonstration of such beneficial effects in various clinical studies has established them as one of the important components of antidiabetic therapy. However, in the light of recent safety concerns raised on such molecules would help prescribers to take an informed decision about risks *versus* benefits while prescribing these agents to their patients.

Keywords: Canagliflozin, Dapagliflozin, Diabetes mellitus, Empagliflozin, Hyperglycemia, Mycotic genital infections, Osmotic diuresis, SGLT-2 inhibitors.

INTRODUCTION

Diabetes mellitus (DM) is currently one of the most prevalent diseases in the world. DM presently possesses a big global disease burden with an estimated prevalence of 422 million cases as per the WHO 2016 data and is likely to be doubled by the year 2030 [1, 2].

* **Corresponding author Pramod Kumar Sharma:** All India Institute of Medical sciences, Jodhpur, Rajasthan, India; E-mail: Pramod309@gmail.com

Atta-ur-Rahman (Ed.)

Type 2 diabetes mellitus (T2DM) is a systemic disorder with characteristic hyperglycemia that results due to β cell dysfunction and/or insulin resistance in the peripheral tissues or both [3]. In the present scenario with an increase in the population, the incidence of the disease is increasing day by day. Factors such as age, urbanization, and a sedentary lifestyle, are usually common risk factors. The strongest risk for hyperglycemia and insulin resistance can be attributed to obesity which can be further possibly related to subclinical inflammation and increased oxidative stress which culminates into damaging the β-cells in the pancreas [4, 5].

Need for New Targets

Antidiabetic drugs in the treatment of T2DM predominantly act by increasing insulin release from the pancreas, increasing the peripheral insulin sensitivity, negatively regulating glucagon secretion, modulating hepatic glucose production or blocking the intestinal glucose uptake [6]. With the progression of disease in T2DM patients, there is a worsening function of the pancreatic β-cells and/or associated with increased insulin resistance, there is perpetually a persistent need for exploring newer drug targets and treatment strategies to control/cure the disease [7]. Several conventional antidiabetics are available, which come with multiple pitfalls, *e.g.,* many sulfonylureas are associated with variable cardiovascular disease (CVD) risk and mortality outcomes [8, 9]. Weight gain is an issue commonly related to thiazolidinediones and sulfonylurea which can further turn out to be a burden in T2DM [10]. In the newly developed agents, Glucagon-like peptide 1 (GLP- 1) analogues are associated with gastroparesis and risk of thyroid cancer in animal models [11, 12]. The side effects of pancreatitis with GLP-1 analogue were initially concerned, but a recent study has confirmed it to be negative [13]. Earlier, Dipeptidyl peptidase 4 (DPP-4) inhibitors, have been reported to cause nasopharyngitis, upper respiratory tract infection, headache and pancreatitis but recent evidence shows there is no risk of increased infections or pancreatitis with these drugs [14, 15]. The limitations of existing therapies serve as a ground for the development of newer molecules which can help improve glycemic control in T2DM without the danger of hypoglycemia, weight gain, improve β-cell function and decrease complications associated with diabetes [16].

Glucose Handling by Kidneys

In normoglycemic individuals having a mean plasma glucose concentration of - 5.5 mmol/L and a normal glomerular filtration rate (GFR) of 125 ml/min/1.73 m^2 in adults, the kidney plays a significant role in glucose homeostasis by reabsorption of about 160–180 g of glucose that kidneys filter each day [17, 18]. A normal individual has a very minute or no glucose in the urine. Kidneys maintain this glucose homeostasis because almost 99% of the filtered glucose gets reabsorbed by the proximal tubule and returns to the blood. Kidneys also contribute to gluconeogenesis and help to maintain the blood glucose level, both the mechanisms are independent of each other [19]. Sustained hyperglycemia as in DM leads to defects in the absorption of glucose which leads to glucosuria. Hyperglycemia enhances the amounts of glucose filtered and increases the reabsorption of glucose.

Role Of SGLT-2 In Kidneys

About 80-90% of the glucose load is reabsorbed from the proximal tubule by the high-capacity sodium-glucose cotransporter 2 (SGLT2). Whereas, remaining 10–20% glucose is absorbed by SGLT1 which is more in the distal parts of the proximal convoluted tubule (PCT) [20].

SGLT-2 Inhibitors- A Targeted Approach to Curb Hyperglycaemia

History

Phlorizin was the first developed SGLT-2 inhibitor, which was extracted from the apple tree bark in 1835. The development of phlorizin was terminated because of its non-selective nature, its rapid degradation by *lactase-phlorizin hydrolase* in the intestine and poor absorption from the gastrointestinal tract led to its low bioavailability and local gastrointestinal (GI) adverse effects like diarrhea [21]. Similarly, sergliflozin also failed in clinical development because of low bioavailability and short half-life [22].

With the increasing importance of this target, continued researches led to the development of novel selective SGLT-2 inhibitors. SGLT-1 selectivity increases GI adverse effects like diarrhea, so to increase the SGLT-2: SGLT-1 selectivity and to minimize the GI adverse effects, the chemical structure of phlorizin was modified [23]. The C-link between the glucose and phenol moiety in phlorizin was replaced by O-link which imparted greater resistance to β-glucosidase and other enzymes leading to greater oral bioavailability of the newer molecules [24]. The molecular modifications increased the SGLT-2 selectivity and half-life [24].

SGLT-2 inhibitors

The mechanism of glucose control in DM is by blockade SGLT-2 inhibition, which leads to an increase in glucose excretion through the urine. As these molecules do not have any effect on insulin secretion, therefore, the risk of hypoglycemia is less, the loss of glucose leads to calorie deficit, which would decrease the body weight, an effect which is instrumental in DM [25, 26].

Among the SGLT-2 inhibitors, Canagliflozin (Fig. **1**) was the first one to be approved in 2013 for use by USFDA in T2DM to improve glycemic control. It is approved to be used along with diet and exercise used in T2DM and not to be used T1DM or in diabetic ketoacidosis [27]. The maximal plasma concentration of canagliflozin is achieved after 1–2 h after oral administration, with an oral bioavailability of 65%. It takes 4–5 days to reach the steady-state concentration. UGT1A9 and UGT2B4 metabolize canagliflozin by glucuronidation [28]. Canagliflozin has been shown to be effective in reducing fasting plasma glucose and HbA1c at a daily dose between 50mg and 600mg [7, 29]. Early intervention and effective glycemic control have been a proven strategy to retard microvascular complications [30]. Canagliflozin should be given to patients with intact renal functions and avoided in patients with eGFR <45 mL/min/1.73 m^2. Besides blocking the renal reabsorption of glucose, it also has partial activity on SGLT1 present in the intestine which might decrease the postprandial glucose absorption [31].

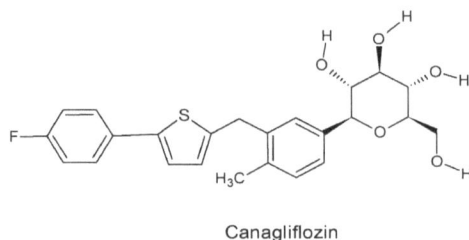

Canagliflozin

Fig. (1). Chemical structure of Canagliflozin.

Dapagliflozin (Fig. **2**) is seen to be far more selective to SGLT2 as compared to SGLT1 [21]. It blocks renal glucose reabsorption to an extent of 40–50%, with glucose excretion up to 80–85 g per day [32]. Trials of dapagliflozin have shown it to be effective as monotherapy or as add-on therapy with oral antidiabetics or insulin in patients with T2DM in reducing fasting and postprandial glucose and HbA1c levels. Literature reveals that high-fat meal does not alter pharmacokinetics and bioavailability of dapagliflozin and lesser interactions with other anti-diabetic drugs are also observed as a desirable property [33, 34].

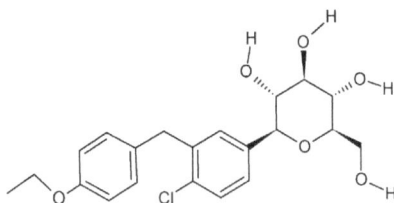

Fig. (2). Chemical structure of Dapagliflozin.

The studies show that Empagliflozin (Fig. **3**) is administered in daily doses of 10 and 25 mg. Trials like Empagliflozin, Cardiovascular Outcomes, and Mortality in Type 2 Diabetes (EMPA-REG OUTCOME) and Canagliflozin and Cardiovascular and Renal Events in Type 2 Diabetes (CANVAS), have shown that SGLT2 inhibitors reduce the risk of cardiovascular events when compared to placebo [35 - 37] Empagliflozin can be administered once daily irrelevant of meal timings either with metformin or as add-on therapy with other anti-hyperglycemic agents including insulin [37]. Reports from the EMPAREG OUTCOME trial revealed many significant effects of empagliflozin. In patients with peripheral artery disease, which is one of the most common complications in patients with type 2 diabetes, it showed the addition of empagliflozin to the standard therapy, reduced the rates of cardiovascular death by 43%, all-cause death by 38%, MACE (major adverse cardiovascular events; cardiovascular death, nonfatal myocardial infarction, or nonfatal stroke) by 16%, hospitalization for heart failure by 44%, and progression of renal dysfunction by 46% when compared to placebo, with no increase in the risk of lower limb amputation [36, 38]. In a renal disease, Empagliflozin should be used with caution and as per the prescribing information by European Medicines Agency (EMA), it should not be given to patients with an eGFR <45 mL/min/1.73 m^2. In patients with an eGFR <60 mL/min/1.73 m2 if on empagliflozin, renal function should be monitored periodically [39, 40]. Most trials have reported a little elevation in HDL-C with no alteration in triglycerides when compared to placebo. Minimal elevation or no change in LDL-C was noted [41 - 43].

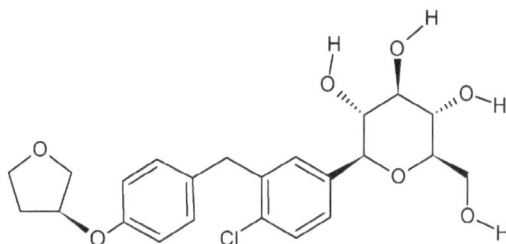

Fig. (3). Chemical structure of Empagliflozin.

Ipragliflozin is an SGLT 2 inhibitor that has been approved in Japan for the treatment of type 2 diabetes for use as monotherapy or in combination with another antihyperglycemic agent. Ipragliflozin has been administered in doses of 25 and 50 mg and the recommended dose is 50 mg once daily, in the morning which can be increased to up to 100 mg once a day [44, 45].

Tofogliflozin is another molecule of the same class which has been approved in Japan in 2014 for the treatment of T2DM either as monotherapy or with other oral anti-diabetic drugs. Tofogliflozin is approved as a 20 mg once daily dose [46].

Luseogliflozin is another molecule of this class which received its approval in Japan for the treatment of T2DM in March 2014. It is approved in the dosage of 2.5 and 5 mg oral tablets once daily [47, 48].

Ertugliflozin is the fourth SGLT2 inhibitor to be approved by the FDA in 2018. It is marketed as a combination with metformin and sitagliptin. The recommended starting dosage of ertugliflozin is 5 mg once daily, which can be increased to a maximum dose of 15 mg per day. In the fixed-dose combinations, the ertugliflozin/metformin combination can be given with a maximum dosage of 7.5/1000 mg twice daily with food [49]. The recommended starting dosage of ertugliflozin/sitagliptin combination is 5/100 mg once daily, with a maximum daily dosage of 15/100 mg. Ertugliflozin should be used with caution in patients with renal injury and should not be used alone or in combination with other antidiabetics in patients with eGFR 30-<60 mL/min/1.73 m^2 and is contraindicated in those with an eGFR <30 mL/min/1.73 m^2 [50, 51].

Sotagliflozin has been developed as a dual inhibitor of SGLT1 and SGLT2. The differentiating clinical feature of dual inhibitor is also blocking gastrointestinal SGLT1 which could prove to be an indispensable therapeutic target to treat diabetes [52]. The advantage of dual inhibition includes a large postprandial glucose reduction, elevation of glucagon-like peptide 1 and modest urinary glucose excretion [53]. Previous studies with genetic and pharmacological inhibition of SGLT1 in the small intestine have shown increased glucose delivery more to the distal gut and colon which in turn induces a sustained increase in circulating GLP-1 secreted by L cells in the distal gut and colon [18, 53]. By delaying intestinal glucose absorption and facilitating renal glucose excretion as well as increasing circulating GLP-1 levels, SGLT1 inhibition appears as an attractive antidiabetic strategy. These characteristics of sotagliflozin seem to have an edge over other previous SGLT 2 inhibitors and hence the trials have shown its utility in the treatment of both type 1 and type 2 diabetes [54]. With some added benefits as compared to other drugs in this class, sotagliflozin has shown to have minor postprandial plasma glucose digression, decreased insulin requirements, suppression of appetite and weight loss which make it a more recommended option as compared to previous ones [55]. Due to its novel mechanism, it has been tried in patients with type 1 diabetes with an adjunct to insulin, which has shown positive results and the molecule is under the review of FDA currently and yet to be approved [56, 57].

Apart from the molecules described above, few of the agents which are in different phases of clinical development are summarized in Table **1**.

Table 1. Summary of drugs in the pipeline.

SGLT-2 Inhibitors	Phase of Development
Bexagliflozin	Phase 3
Remogliflozin	Phase 3

GW 869682, ISIS 388626 and GSK-1614235 are few molecules that are in the early phase of drug development.

SGLT-2 Inhibitors in Renal Failure

Almost all the SGLT-2 inhibitors should be used with caution in patients with renal disease and contraindicated in patients with end-stage renal disease. With the continuous decrease in GFR, there is also a reduction in the amount of filtered glucose in PCT. Hence the glucose-lowering efficacy of SGLT2 inhibitors decreases with the decline in GFR [18].

Safety and Tolerability of SGLT-2 Inhibitors

Diabetes is a kind of metabolic disorder that requires lifelong treatment, and hence the safety of the drug to be given holds utmost importance. There is an increased amount of glucose excretion in the urine, which is explained by its mechanism of action leading to the formation of a conducive environment for microbial growth. This leads to higher rates of genital infections and urinary tract infections (UTIs) in patients given SGLT-2 inhibitors [58 - 66]. The incidence of UTI was found to be more in females as compared to males [67, 68]. As the mechanism of action of SGLT-2 inhibitors is insulin-independent, the risk of hypoglycemia seems to be minimal. However, when these drugs are given in combination with other anti-diabetic drugs, the risk of hypoglycemia increases [58, 60 - 64, 69]. There is an increased incidence of genital mycotic infections usually mild to moderate in severity which may require treatment with antifungals [68]. In studies of canagliflozin, the uncircumcised males were seemed to be at a higher risk of developing genital infections like balanitis or balanoposthitis [70]. Similar genital mycotic infections were also seen in females like vulvitis, vulvovaginal candidiasis, vulvovaginitis [66, 68, 71]. The literature shows there is a mild elevation of the hematocrit of about 1–2% observed in with the administration of SGLT-2 inhibitors along with mild volume contraction. The urine volume rises, but modestly due to osmotic diuresis while there is no electrolyte imbalance seen [72, 73]. Osmotic diuresis occasionally leads to transient hypotensive episodes secondary to volume reduction which can culminate into acute kidney injury [74]. There is a slight increase in serum creatinine with the SGLT-2 inhibitors and with a substantial increase seen with canagliflozin [75]. The estimated glomerular filtration rate (eGFR) has been seen to decrease proportionally with a moderate increase in blood urea nitrogen (BUN) predominantly in patients with chronic kidney disease [76, 77]. There have been reports of Diabetic Ketoacidosis with SGLT2 inhibitors that may be increased among patients with insulin-deficient diabetes (T1DM or including those with long term T2DM) which is further escalated by severe acute illness, dehydration, stress, surgery or excessive alcohol intake [78, 79]. Canagliflozin, as per the previous trials was thought to be associated with a decreased total hip bone mineral density, which could increase the fracture risk [80, 81]. However, recent evidence and few meta-analyses have shown there is no significant fracture risk with SGLT-2 inhibitors when compared to placebo [82 - 85]. The CANVAS (Canagliflozin Cardiovascular Assessment Study) and CANVAS-R (A Study of the Effects of Canagliflozin on Renal Endpoints in Adult Participants With Type 2 Diabetes Mellitus) have shown that leg and foot amputations were more prevalent in patients treated with canagliflozin when compared to the patients treated with placebo [86]. Recently, FDA has issued warning against SGLT-2 inhibitors regarding the risk of rare but serious infection of the genitals and the area around the genitals, called necrotizing fasciitis of the perineum, which is also referred to as Fournier's gangrene [87].

Beneficial Effects of SGLT-2 Inhibitors Beyond Controlling Hyperglycaemia

Their role in reducing plasma fasting glucose and achieving glycaemic control has been well established. Apart from achieving target HbA1c, they have also been seen to reduce bodyweight with the largest weight reduction associated with canagliflozin [88, 89]. Osmotic diuresis can be a reason for early weight loss and sustained effect of weight loss over a long time can be attributed to caloric loss [88 - 90]. The decrease in body weight observed with both canagliflozin and dapagliflozin use was predominate because of loss of fat mass rather than lean body mass and the lost fat was marginally more from visceral abdominal tissue as compared to subcutaneous abdominal tissue with slimming of waist circumference, may be associated with improved insulin sensitivity [91, 92].

Most studies of SGLT2 inhibitors have shown a significant reduction in blood pressure, with greater reductions seen in the systolic component as compared to diastolic [58, 68, 93 - 95]. The effects on blood pressure were not found to be dose-dependent and no substantial effect on heart rate was seen nor there were increased episodes of hypotension or syncope [95]. The initial depletion of blood pressure observed may be attributed to osmotic diuresis caused by the inhibition of renal glucose and sodium reabsorption [96]. The sustained control of blood pressure can be theoretically explained by local inhibition of the renin-angiotensin system (RAS). SGLT2 blockade in the proximal convoluted tubule by its inhibitors results in an increase in the sodium load in the fluid passing through the distal convoluted tubule [97]. The juxtaglomerular apparatus contains macula densa cell that senses the increase in sodium load and results in the inhibition of renin release from the juxtaglomerular cells [97]. This mechanism of RAS inhibition explains the reduction in blood pressure to a greater extent yet, with chronic use of SGLT2 inhibitors leads to a diminution of sodium excretion due to tubuloglomerular feedback [98]. Yet, these drugs have shown a cardiovascular benefit in patients with diabetes as a sustained, controlled blood pressure, which proves to be an edge in patients of diabetes with hypertension [99].

The effects of SGLT2 inhibitors on lipid profile were found to be highly variable. Canagliflozin had a favorable effect on the lipid profile with an elevation of high-density lipoprotein, a decrease in triglyceride and an increase in low-density lipoprotein [68, 95]. Whereas, dapagliflozin was found to be lipid neutral and no such effect was demonstrated [100, 101].

Trials like EMPA-REG OUTCOME, which studied empagliflozin, has shown to reduce cardiovascular mortality and heart failure in high-risk patients (ischemic heart disease, stroke and/or heart failure) with Type 2 DM with a previous history of cardiovascular event or with an established cardiac problem [102]. The trial showed a reduction in CV death, all-cause mortality, reduction in hospitalizations and the effects were identical with both doses of empagliflozin, 10 mg or 25 mg once daily [103].

The CANVAS-PROGRAM study, which was performed with canagliflozin concluded that the cardiovascular benefits which were evident with empagliflozin are not limited to it but can also be demonstrated with canagliflozin. However, all the patients included in the CANVAS trial did not establish cardiovascular disease as it was in EMPA-REG OUTCOME. Still the deaths associated with cardiovascular diseases, nonfatal MI, or nonfatal stroke, were significantly lower with canagliflozin as compared to placebo [104].

Another recent trial, Dapagliflozin Effect on Cardiovascular Events (DECLARE)-Thrombolysis in Myocardial Infarction (TIMI)-58 showed that in the T2DM patients with who had a previous episode or were at risk for atherosclerotic cardiovascular disease, when treated with dapagliflozin did not alter the incidence of MACE but, did result in lower death rate due to cardiovascular events or hospitalization due to heart failure [105].

It has been postulated that the cardiovascular benefits with this class of drug can also be due to its effect on modifying various risk factors such as lowering blood glucose, blood pressure, maintaining arterial elasticity, decreasing weight, and visceral adipose tissue deposition, altering lipid parameters and decreasing albuminuria and serum uric acid levels [102].

Evidence shows they also ameliorate oxidative stress and shift in fuel metabolism in the heart from fatty acids to beta-hydroxybutyrate. The potential hemodynamic advantages, which could further help to decrease cardiovascular risks, are decreased pre- and afterload enhanced myocardial oxygen supply and enhanced cardiac systolic and diastolic function [102, 106]. Some of the evidence suggest empagliflozin minimizes sodium and calcium overload in the myocardial cell and increases the concentration of calcium in the myocardial mitochondria and hence reduces the probability of the development of heart failure [107].

CONCLUSION

The evidence so far accumulated suggests that SGLT-2 inhibitors not only help reduce blood glucose but also have a myriad of beneficial effects independent of controlling serum glucose. They have also proven to improve survival by decreasing the incidence of cardiovascular events; a big advantage as the patients of diabetes are always at increased risk of developing cardiovascular events. Considering the benefits offered by SGLT-2 inhibitors over existing antidiabetics, they hold a promising place in the therapeutic armamentarium of diabetes management. However, the recent safety warning attached to these molecules would necessitate healthcare providers to balance risk *versus* benefit while treating a patient with SGLT-2 inhibitors.

CONSENT FOR PUBLICATION

Not applicable.

CONFLICT OF INTEREST

The author confirms that he has no conflict of interest to declare for this publication.

ACKNOWLEDGEMENTS

We acknowledge Dr. Surjit Singh for assisting in the literature search. The chemical structures were adapted from Pubchem and drawn by ACD/ChemSketch Freeware software.

REFERENCES

[1]   World Health Organization. Global Report on Diabetes 2016. https://apps.who.int/ iris/bitstream/handle/10665/204871/9789241565257_eng.pdf;jsessionid=59263CA281AE3D0EA62F EF8967B0ABAF?sequence=1

[2]   Wild S, Roglic G, Green A, Sicree R, King H. Global prevalence of diabetes: estimates for the year

2000 and projections for 2030. Diabetes Care 2004; 27(5): 1047-53.
[http://dx.doi.org/10.2337/diacare.27.5.1047] [PMID: 15111519]

[3] Defronzo RA. Banting Lecture. From the triumvirate to the ominous octet: a new paradigm for the treatment of type 2 diabetes mellitus. Diabetes 2009; 58(4): 773-95.
[http://dx.doi.org/10.2337/db09-9028] [PMID: 19336687]

[4] Schwanstecher C, Schwanstecher M. Targeting type 2 diabetes. Handb Exp Pharmacol 2011; 203(203): 1-33.
[PMID: 21484565]

[5] Donath MY, Shoelson SE. Type 2 diabetes as an inflammatory disease. Nat Rev Immunol 2011; 11(2): 98-107.
[http://dx.doi.org/10.1038/nri2925] [PMID: 21233852]

[6] Wagman AS, Nuss JM. Current therapies and emerging targets for the treatment of diabetes. Curr Pharm Des 2001; 7(6): 417-50.
[http://dx.doi.org/10.2174/1381612013397915] [PMID: 11281851]

[7] Kaushal S, Singh H, Thangaraju P, Singh J. Canagliflozin: A Novel SGLT2 Inhibitor for Type 2 Diabetes Mellitus. N Am J Med Sci 2014; 6(3): 107-13.
[http://dx.doi.org/10.4103/1947-2714.128471] [PMID: 24741548]

[8] Schramm TK, Gislason GH, Vaag A, *et al.* Mortality and cardiovascular risk associated with different insulin secretagogues compared with metformin in type 2 diabetes, with or without a previous myocardial infarction: a nationwide study. Eur Heart J 2011; 32(15): 1900-8.
[http://dx.doi.org/10.1093/eurheartj/ehr077] [PMID: 21471135]

[9] Gore MO, McGuire DK. Resolving drug effects from class effects among drugs for type 2 diabetes mellitus: more support for cardiovascular outcome assessments. Eur Heart J 2011; 32(15): 1832-4.
[http://dx.doi.org/10.1093/eurheartj/ehr019] [PMID: 21471136]

[10] Wilding J. Thiazolidinediones, insulin resistance and obesity: Finding a balance. Int J Clin Pract 2006; 60(10): 1272-80.
[http://dx.doi.org/10.1111/j.1742-1241.2006.01128.x] [PMID: 16981971]

[11] Gore MO, McGuire DK. Resolving drug effects from class effects among drugs for type 2 diabetes mellitus: more support for cardiovascular outcome assessments. Eur Heart J 2011; 32(15): 1832-4.
[http://dx.doi.org/10.1093/eurheartj/ehr019] [PMID: 21471136]

[12] Zinman B, Schmidt WE, Moses A, Lund N, Gough S. Achieving a clinically relevant composite outcome of an HbA1c of <7% without weight gain or hypoglycaemia in type 2 diabetes: a meta-analysis of the liraglutide clinical trial programme. Diabetes Obes Metab 2012; 14(1): 77-82.
[http://dx.doi.org/10.1111/j.1463-1326.2011.01493.x] [PMID: 21883806]

[13] Azoulay L, Filion KB, Platt RW, *et al.* Association Between Incretin-Based Drugs and the Risk of Acute Pancreatitis. JAMA Intern Med 2016; 176(10): 1464-73.
[http://dx.doi.org/10.1001/jamainternmed.2016.1522] [PMID: 27479930]

[14] Januvia (sitagliptin) . https://www.accessdata.fda.gov/drugsatfda_docs/label/2012/021995s019lbl.pdf

[15] Yang W, Cai X, Han X, Ji L. DPP-4 inhibitors and risk of infections: a meta-analysis of randomized controlled trials. Diabetes Metab Res Rev 2016; 32(4): 391-404.
[http://dx.doi.org/10.1002/dmrr.2723] [PMID: 26417956]

[16] Mathers MC, Zarbock SD, Sutton EE. New and Future Medications for the Treatment of Type 2 Diabetes Mellitus. Am J Lifestyle Med 2014; 8(4): 260-80.
[http://dx.doi.org/10.1177/1559827613498694]

[17] Mather A, Pollock C. Glucose handling by the kidney. Kidney Int Suppl 2011; (120): S1-6.
[http://dx.doi.org/10.1038/ki.2010.509] [PMID: 21358696]

[18] Vallon V. The mechanisms and therapeutic potential of SGLT2 inhibitors in diabetes mellitus. Annu

Rev Med 2015; 66: 255-70.
[http://dx.doi.org/10.1146/annurev-med-051013-110046] [PMID: 25341005]

[19]   Gerich JE, Meyer C, Woerle HJ, Stumvoll M. Renal gluconeogenesis: its importance in human glucose homeostasis. Diabetes Care 2001; 24(2): 382-91.
[http://dx.doi.org/10.2337/diacare.24.2.382] [PMID: 11213896]

[20]   Wright EM, Loo DD, Hirayama BA. Biology of human sodium glucose transporters. Physiol Rev 2011; 91(2): 733-94.
[http://dx.doi.org/10.1152/physrev.00055.2009] [PMID: 21527736]

[21]   Ferrannini E, Solini A. SGLT2 inhibition in diabetes mellitus: rationale and clinical prospects. Nat Rev Endocrinol 2012; 8(8): 495-502.
[http://dx.doi.org/10.1038/nrendo.2011.243] [PMID: 22310849]

[22]   Hussey EK, Dobbins RL, Stoltz RR, *et al.* Multiple-dose pharmacokinetics and pharmacodynamics of sergliflozin etabonate, a novel inhibitor of glucose reabsorption, in healthy overweight and obese subjects: a randomized double-blind study. J Clin Pharmacol 2010; 50(6): 636-46.
[http://dx.doi.org/10.1177/0091270009352185] [PMID: 20200268]

[23]   Rossetti L, Smith D, Shulman GI, Papachristou D, DeFronzo RA. Correction of hyperglycemia with phlorizin normalizes tissue sensitivity to insulin in diabetic rats. J Clin Invest 1987; 79(5): 1510-5.
[http://dx.doi.org/10.1172/JCI112981] [PMID: 3571496]

[24]   Washburn WN, Poucher SM. Differentiating sodium-glucose co-transporter-2 inhibitors in development for the treatment of type 2 diabetes mellitus. Expert Opin Investig Drugs 2013; 22(4): 463-86.
[http://dx.doi.org/10.1517/13543784.2013.774372] [PMID: 23452053]

[25]   List JF, Whaley JM. Glucose dynamics and mechanistic implications of SGLT2 inhibitors in animals and humans. Kidney Int Suppl 2011; 79(120): S20-7.
[http://dx.doi.org/10.1038/ki.2010.512] [PMID: 21358698]

[26]   Bolinder J, Ljunggren Ö, Kullberg J, *et al.* Effects of dapagliflozin on body weight, total fat mass, and regional adipose tissue distribution in patients with type 2 diabetes mellitus with inadequate glycemic control on metformin. J Clin Endocrinol Metab 2012; 97(3): 1020-31.
[http://dx.doi.org/10.1210/jc.2011-2260] [PMID: 22238392]

[27]   INVOKANA™ (Canagliflozin) for the Treatment of Adults with Type 2 Diabetes. https://www. jnj.com/media-center/press-releases/us-fda-approves-invokana-canagliflozin-for-the-treatment-of-adults-with-type-2-diabetes

[28]   Kasichayanula S, Liu X, Lacreta F, Griffen SC, Boulton DW. Clinical pharmacokinetics and pharmacodynamics of dapagliflozin, a selective inhibitor of sodium-glucose co-transporter type 2. Clin Pharmacokinet 2014; 53(1): 17-27.
[http://dx.doi.org/10.1007/s40262-013-0104-3] [PMID: 24105299]

[29]   Rosenstock J, Aggarwal N, Polidori D, *et al.* Dose-ranging effects of canagliflozin, a sodium-glucose cotransporter 2 inhibitor, as add-on to metformin in subjects with type 2 diabetes. Diabetes Care 2012; 35(6): 1232-8.
[http://dx.doi.org/10.2337/dc11-1926] [PMID: 22492586]

[30]   Holman RR, Paul SK, Bethel MA, Matthews DR, Neil HA. 10-year follow-up of intensive glucose control in type 2 diabetes. N Engl J Med 2008; 359(15): 1577-89.
[http://dx.doi.org/10.1056/NEJMoa0806470] [PMID: 18784090]

[31]   Polidori D, Sha S, Mudaliar S, *et al.* Canagliflozin lowers postprandial glucose and insulin by delaying intestinal glucose absorption in addition to increasing urinary glucose excretion: results of a randomized, placebo-controlled study. Diabetes Care 2013; 36(8): 2154-61.
[http://dx.doi.org/10.2337/dc12-2391] [PMID: 23412078]

[32]   Meng W, Ellsworth BA, Nirschl AA, *et al.* Discovery of dapagliflozin: a potent, selective renal

sodium-dependent glucose cotransporter 2 (SGLT2) inhibitor for the treatment of type 2 diabetes. J Med Chem 2008; 51(5): 1145-9.
[http://dx.doi.org/10.1021/jm701272q] [PMID: 18260618]

[33]   Kasichayanula S, Liu X, Zhang W, *et al.* Effect of a high-fat meal on the pharmacokinetics of dapagliflozin, a selective SGLT2 inhibitor, in healthy subjects. Diabetes Obes Metab 2011; 13(8): 770-3.
[http://dx.doi.org/10.1111/j.1463-1326.2011.01397.x] [PMID: 21435141]

[34]   Kasichayanula S, Liu X, Shyu WC, *et al.* Lack of pharmacokinetic interaction between dapagliflozin, a novel sodium-glucose transporter 2 inhibitor, and metformin, pioglitazone, glimepiride or sitagliptin in healthy subjects. Diabetes Obes Metab 2011; 13(1): 47-54.
[http://dx.doi.org/10.1111/j.1463-1326.2010.01314.x] [PMID: 21114603]

[35]   Mahaffey KW, Neal B, Perkovic V, *et al.* Canagliflozin for primary and secondary prevention of cardiovascular events: results from the CANVAS program (Canagliflozin Cardiovascular Assessment Study). Circulation 2018; 137(4): 323-34.
[http://dx.doi.org/10.1161/CIRCULATIONAHA.117.032038] [PMID: 29133604]

[36]   Verma S, Mazer CD, Al-Omran M, *et al.* Cardiovascular Outcomes and Safety of Empagliflozin in Patients With Type 2 Diabetes Mellitus and Peripheral Artery Disease: A Subanalysis of EMPA-REG OUTCOME. Circulation 2018; 137(4): 405-7.
[http://dx.doi.org/10.1161/CIRCULATIONAHA.117.032031] [PMID: 29133602]

[37]   Fujita Y, Inagaki N. Renal sodium glucose cotransporter 2 inhibitors as a novel therapeutic approach to treatment of type 2 diabetes: Clinical data and mechanism of action. J Diabetes Investig 2014; 5(3): 265-75.
[http://dx.doi.org/10.1111/jdi.12214] [PMID: 24843771]

[38]   Fernández-Ruiz I. Diabetes: Further insights into SGLT2 inhibitors. Nat Rev Cardiol 2018; 15(1): 2-3.
[http://dx.doi.org/10.1038/nrcardio.2017.198] [PMID: 29188808]

[39]   Neumiller JJ. Empagliflozin: a new sodium-glucose co-transporter 2 (SGLT2) inhibitor for the treatment of type 2 diabetes. Drugs Context 2014; 3212262
[PMID: 24991224]

[40]   Jardiance® (empagliflozin). Full Prescribing Information, Boehringer Ingelheim Pharmaceuticals and Eli Lilly and Company, Ingelheim, Germany, and Indianapolis,IN, USA 2014.https://docs.boehringer-ingelheim.com/Prescribing%20Information/PIs/Jardiance/jardiance.pdf

[41]   Roden M, Weng J, Eilbracht J, *et al.* Empagliflozin monotherapy with sitagliptin as an active comparator in patients with type 2 diabetes: a randomised, double-blind, placebo-controlled, phase 3 trial. Lancet Diabetes Endocrinol 2013; 1(3): 208-19.
[http://dx.doi.org/10.1016/S2213-8587(13)70084-6] [PMID: 24622369]

[42]   Häring H-U, Merker L, Seewaldt-Becker E, *et al.* EMPA-REG METSU Trial Investigators. Empagliflozin as add-on to metformin plus sulfonylurea in patients with type 2 diabetes: a 24-week, randomized, double-blind, placebo-controlled trial. Diabetes Care 2013; 36(11): 3396-404.
[http://dx.doi.org/10.2337/dc12-2673] [PMID: 23963895]

[43]   Ridderstråle M, Andersen KR, Zeller C, Kim G, Woerle HJ, Broedl UC. EMPA-REG H2H-SU trial investigators. Comparison of empagliflozin and glimepiride as add-on to metformin in patients with type 2 diabetes: a 104-week randomised, active-controlled, double-blind, phase 3 trial. Lancet Diabetes Endocrinol 2014; 2(9): 691-700.
[http://dx.doi.org/10.1016/S2213-8587(14)70120-2] [PMID: 24948511]

[44]   https://www.astellas.com/en/corporate/news/pdf/140117_1_Eg.pdf

[45]   Poole RM, Dungo RT. Ipragliflozin: first global approval. Drugs 2014; 74(5): 611-7.
[http://dx.doi.org/10.1007/s40265-014-0204-x] [PMID: 24668021]

[46]   Poole RM, Prossler JE. Tofogliflozin: first global approval. Drugs 2014; 74(8): 939-44.

[http://dx.doi.org/10.1007/s40265-014-0229-1] [PMID: 24848755]

[47] Markham A, Elkinson S. Luseogliflozin: first global approval. Drugs 2014; 74(8): 945-50.
[http://dx.doi.org/10.1007/s40265-014-0230-8] [PMID: 24848756]

[48] http://medical.taishotoyama.co.jp/cgi-bin/index2_n.cgi?mode=sel&rows=86&caut=tenp&page=/data/tenp/pdf/tenp_lsf.pdf

[49] https://www.ema.europa.eu/documents/product-information/steglatro-epar-product-information_en.pdf

[50] https://www.pfizer.com/news/press-release/press-release-detail/fda_approves_sglt2_inhibitor_steglatro_ertugliflozin_and_fixed_dose_combination_steglujan_ertugliflozin_and_sitagliptin_for_adults_with_type_2_diabetes

[51] Ertugliflozin for Type 2 Diabetes. JAMA 2018; 319(23): 2434-5.
[http://dx.doi.org/10.1001/jama.2018.5840] [PMID: 29922825]

[52] Lapuerta P, Zambrowicz B, Strumph P, Sands A. Development of sotagliflozin, a dual sodium-dependent glucose transporter 1/2 inhibitor. Diab Vasc Dis Res 2015; 12(2): 101-10.
[http://dx.doi.org/10.1177/1479164114563304] [PMID: 25690134]

[53] Zambrowicz B, Ogbaa I, Frazier K, *et al.* Effects of LX4211, a dual sodium-dependent glucose cotransporters 1 and 2 inhibitor, on postprandial glucose, insulin, glucagon-like peptide 1, and peptide tyrosine tyrosine in a dose-timing study in healthy subjects. Clin Ther 2013; 35(8): 1162-1173.e8.
[http://dx.doi.org/10.1016/j.clinthera.2013.06.011] [PMID: 23911260]

[54] Zambrowicz B, Freiman J, Brown PM, *et al.* LX4211, a dual SGLT1/SGLT2 inhibitor, improved glycemic control in patients with type 2 diabetes in a randomized, placebo-controlled trial. Clin Pharmacol Ther 2012; 92(2): 158-69.
[http://dx.doi.org/10.1038/clpt.2012.58] [PMID: 22739142]

[55] Sims H, Smith KH, Bramlage P, Minguet J. Sotagliflozin: a dual sodium-glucose co-transporter-1 and -2 inhibitor for the management of Type 1 and Type 2 diabetes mellitus. Diabet Med 2018; 35(8): 1037-48.
[http://dx.doi.org/10.1111/dme.13645] [PMID: 29637608]

[56] Garg SK, Strumph P. Effects of Sotagliflozin Added to Insulin in Type 1 Diabetes. N Engl J Med 2018; 378(10): 967-8.
[PMID: 29514037]

[57] http://www.news.sanofi.us/2018-05-22-FDA-to-review-Zynquista-TM-sotagliflozin-as-potential-treatment-for-type-1-diabetes

[58] Clar C, Gill JA, Court R, Waugh N. Systematic review of SGLT2 receptor inhibitors in dual or triple therapy in type 2 diabetes. BMJ Open 2012; 2(5)e001007
[http://dx.doi.org/10.1136/bmjopen-2012-001007] [PMID: 23087012]

[59] Ferrannini E, Ramos SJ, Salsali A, Tang W, List JF. Dapagliflozin monotherapy in type 2 diabetic patients with inadequate glycemic control by diet and exercise: a randomized, double-blind, placebo-controlled, phase 3 trial. Diabetes Care 2010; 33(10): 2217-24.
[http://dx.doi.org/10.2337/dc10-0612] [PMID: 20566676]

[60] Kawano H, Kashiwagi A, Kazuta K, Yoshida S, Ueyama E, Utsuno A. Longterm safety, tolerability and efficacy of ipragliflozin in Japanese patients with type 2 diabetes mellitus: IGNITE [Abstract] Diabetes 2012; 61 (Suppl. 1) A611 (2422-PO)

[61] Stenlöf K, Cefalu WT, Kim KA, *et al.* Efficacy and safety of canagliflozin monotherapy in subjects with type 2 diabetes mellitus inadequately controlled with diet and exercise. Diabetes Obes Metab 2013; 15(4): 372-82.
[http://dx.doi.org/10.1111/dom.12054] [PMID: 23279307]

[62] List JF, Woo V, Morales E, Tang W, Fiedorek FT. Sodium-glucose cotransport inhibition with dapagliflozin in type 2 diabetes. Diabetes Care 2009; 32(4): 650-7.
[http://dx.doi.org/10.2337/dc08-1863] [PMID: 19114612]

[63] Woerle HJ, Ferrannini E, Berk A, Manun'Ebo M, Pinnetti S, Broedl UC. Safety and efficacy of empagliflozin as monotherapy or add-on to metformin in a 78-week open label extension study in patients with type 2 diabetes [Abstract] Diabetes 2012; 61 (Suppl.1A) LB13 (49-LB)

[64] Bailey CJ, Gross JL, Hennicken D, Iqbal N, Mansfield TA, List JF. Dapagliflozin add-on to metformin in type 2 diabetes inadequately controlled with metformin: a randomized, double-blind, placebo-controlled 102-week trial. BMC Med 2013; 11(43): 43.
[http://dx.doi.org/10.1186/1741-7015-11-43] [PMID: 23425012]

[65] Nauck MA, Del Prato S, Meier JJ, *et al.* Dapagliflozin *versus* glipizide as add-on therapy in patients with type 2 diabetes who have inadequate glycemic control with metformin: a randomized, 52-week, double-blind, active-controlled noninferiority trial. Diabetes Care 2011; 34(9): 2015-22.
[http://dx.doi.org/10.2337/dc11-0606] [PMID: 21816980]

[66] Wilding JP, Woo V, Soler NG, *et al.* Dapagliflozin 006 Study Group. Long-term efficacy of dapagliflozin in patients with type 2 diabetes mellitus receiving high doses of insulin: a randomized trial. Ann Intern Med 2012; 156(6): 405-15.
[http://dx.doi.org/10.7326/0003-4819-156-6-201203200-00003] [PMID: 22431673]

[67] Nauck MA, Del Prato S, Durán-García S, *et al.* Durability of glycaemic efficacy over 2 years with dapagliflozin *versus* glipizide as add-on therapies in patients whose type 2 diabetes mellitus is inadequately controlled with metformin. Diabetes Obes Metab 2014; 16(11): 1111-20.
[http://dx.doi.org/10.1111/dom.12327] [PMID: 24919526]

[68] Rosenwasser RF, Sultan S, Sutton D, Choksi R, Epstein BJ. SGLT-2 inhibitors and their potential in the treatment of diabetes. Diabetes Metab Syndr Obes 2013; 6: 453-67.
[PMID: 24348059]

[69] Kalra S. Sodium Glucose Co-Transporter-2 (SGLT2) Inhibitors: A Review of Their Basic and Clinical Pharmacology. Diabetes Ther 2014; 5(2): 355-66.
[http://dx.doi.org/10.1007/s13300-014-0089-4] [PMID: 25424969]

[70] Unnikrishnan AG, Kalra S, Purandare V, Vasnawala H. Genital infections with sodium glucose cotransporter-2 inhibitors: Occurrence and management in patients with type 2 diabetes mellitus. Indian J Endocrinol Metab 2018; 22(6): 837-42.
[http://dx.doi.org/10.4103/ijem.IJEM_159_17] [PMID: 30766827]

[71] Forst T, Guthrie R, Goldenberg R, *et al.* Efficacy and safety of canagliflozin over 52 weeks in patients with type 2 diabetes on background metformin and pioglitazone. Diabetes Obes Metab 2014; 16(5): 467-77.
[http://dx.doi.org/10.1111/dom.12273] [PMID: 24528605]

[72] Ferrannini E. Sodium-glucose transporter-2 inhibition as an antidiabetic therapy. Nephrol Dial Transplant 2010; 25(7): 2041-3.
[http://dx.doi.org/10.1093/ndt/gfq249] [PMID: 20466683]

[73] Chao EC. SGLT-2 Inhibitors: A New Mechanism for Glycemic Control. Clin Diabetes 2014; 32(1): 4-11.
[http://dx.doi.org/10.2337/diaclin.32.1.4] [PMID: 26246672]

[74] Liu YL, Prowle J, Licari E, Uchino S, Bellomo R. Changes in blood pressure before the development of nosocomial acute kidney injury. Nephrol Dial Transplant 2009; 24(2): 504-11.
[http://dx.doi.org/10.1093/ndt/gfn490] [PMID: 18768582]

[75] Storgaard H, Gluud LL, Bennett C, *et al.* Benefits and Harms of Sodium-Glucose Co-Transporter 2 Inhibitors in Patients with Type 2 Diabetes: A Systematic Review and Meta-Analysis. PLoS One 2016; 11(11)e0166125

[http://dx.doi.org/10.1371/journal.pone.0166125] [PMID: 27835680]

[76]   Kohan DE, Fioretto P, Tang W, List JF. Long-term study of patients with type 2 diabetes and moderate renal impairment shows that dapagliflozin reduces weight and blood pressure but does not improve glycemic control. Kidney Int 2014; 85(4): 962-71.
[http://dx.doi.org/10.1038/ki.2013.356] [PMID: 24067431]

[77]   Hinnen D. Glucuretic effects and renal safety of dapagliflozin in patients with type 2 diabetes. Ther Adv Endocrinol Metab 2015; 6(3): 92-102.
[http://dx.doi.org/10.1177/2042018815575273] [PMID: 26137213]

[78]   Goldenberg RM, Berard LD, Cheng AY, *et al.* SGLT2 Inhibitor associated Diabetic Ketoacidosis: Clinical Review and Recommendations for Prevention and Diagnosis Clin Ther 2016; 38: 2654-64.

[79]   Burke KR, Schumacher CA, Harpe SE. SGLT2 Inhibitors: A Systematic Review of Diabetic Ketoacidosis and Related Risk Factors in the Primary Literature. Pharmacotherapy 2017; 37(2): 187-94.
[http://dx.doi.org/10.1002/phar.1881] [PMID: 27931088]

[80]   Taylor SI, Blau JE, Rother KI. Possible adverse effects of SGLT2 inhibitors on bone. Lancet Diabetes Endocrinol 2015; 3(1): 8-10.
[http://dx.doi.org/10.1016/S2213-8587(14)70227-X] [PMID: 25523498]

[81]   Bilezikian JP, Watts NB, Usiskin K, *et al.* Evaluation of Bone Mineral Density and Bone Biomarkers in Patients With Type 2 Diabetes Treated With Canagliflozin. J Clin Endocrinol Metab 2016; 101(1): 44-51.
[http://dx.doi.org/10.1210/jc.2015-1860] [PMID: 26580234]

[82]   Tang HL, Li DD, Zhang JJ, *et al.* Lack of evidence for a harmful effect of sodium-glucose co-transporter 2 (SGLT2) inhibitors on fracture risk among type 2 diabetes patients: a network and cumulative meta-analysis of randomized controlled trials. Diabetes Obes Metab 2016; 18(12): 1199-206.
[http://dx.doi.org/10.1111/dom.12742] [PMID: 27407013]

[83]   Ruanpeng D, Ungprasert P, Sangtian J, Harindhanavudhi T. Sodium-glucose cotransporter 2 (SGLT2) inhibitors and fracture risk in patients with type 2 diabetes mellitus: A meta-analysis. Diabetes Metab Res Rev 2017; 33(6): 1-11.
[http://dx.doi.org/10.1002/dmrr.2903] [PMID: 28440590]

[84]   Azharuddin M, Adil M, Ghosh P, Sharma M. Sodium-glucose cotransporter 2 inhibitors and fracture risk in patients with type 2 diabetes mellitus: A systematic literature review and Bayesian network meta-analysis of randomized controlled trials. Diabetes Res Clin Pract 2018; 146: 180-90.
[http://dx.doi.org/10.1016/j.diabres.2018.10.019] [PMID: 30389620]

[85]   Li X, Li T, Cheng Y, *et al.* Effects of SGLT2 inhibitors on fractures and bone mineral density in type 2 diabetes: An updated meta-analysis. Diabetes Metab Res Rev 2019; 35(7)e3170
[http://dx.doi.org/10.1002/dmrr.3170] [PMID: 30983141]

[86]   FDA Drug Safety Communication. https://www.fda.gov/Drugs/DrugSafety/ucm557507.htm

[87]   https://www.fda.gov/Drugs/DrugSafety/ucm617360.htm

[88]   Inagaki N, Kondo K, Iwasaki T. Canagliflozin, a novel inhibitor of sodium glucose co-transporter 2(SGLT2) improves glycemic control and reduces body weight in Japanese type 2 diabetes mellitus(T2DM) Diabetes 2011; 60 abstract: 0999-P.

[89]   Rosenstock J, Arbit D, Usiskin K, Capuano G, Canovatchel W. Orlando: American Diabetes Association; 2010 Canagliflozin, an inhibitor of sodium glucose co-transporter 2 (SGLT2), improves glycemic control and lowers body weight in subjects with type 2 diabetes (T2D) on metformin Abstract 77. 2010.

[90]   Rosenstock J, Polidori D, Zhao Y, Sha S, Arbit D, Usiskin K, *et al.* Canagliflozin, an inhibitor of sodium glucose co-transporter 2 (SGLT2), improves glycemic control, lowers body weight and

improves beta cell function in subjects with type 2 diabetes on back- ground metformin (T2D) on metformin. Diabetologia 2010; 53: S1-S556.

[91]   Toubro S, Cefalu WT, Xie J, *et al.* Canagliflozin, a sodium glucose co-transporter 2 inhibitor, reduces body weight mainly through loss of fat mass in subjects with type 2 diabetes. Diabetologia 2012; 55 (Suppl. 1): S313-4. [Abstract].

[92]   Bolinder J, Ljunggren Ö, Kullberg J, *et al.* Effects of dapagliflozin on body weight, total fat mass, and regional adipose tissue distribution in patients with type 2 diabetes mellitus with inadequate glycemic control on metformin. J Clin Endocrinol Metab 2012; 97(3): 1020-31.
[http://dx.doi.org/10.1210/jc.2011-2260] [PMID: 22238392]

[93]   Baker WL, Smyth LR, Riche DM, Bourret EM, Chamberlin KW, White WB. Effects of sodium-glucose co-transporter 2 inhibitors on blood pressure: a systematic review and meta-analysis. J Am Soc Hypertens 2014; 8(4): 262-75.e9.
[http://dx.doi.org/10.1016/j.jash.2014.01.007] [PMID: 24602971]

[94]   Oliva RV, Bakris GL. Blood pressure effects of sodium-glucose co-transport 2 (SGLT2) inhibitors. J Am Soc Hypertens 2014; 8(5): 330-9.
[http://dx.doi.org/10.1016/j.jash.2014.02.003] [PMID: 24631482]

[95]   Foote C, Perkovic V, Neal B. Effects of SGLT2 inhibitors on cardiovascular outcomes. Diab Vasc Dis Res 2012; 9(2): 117-23.
[http://dx.doi.org/10.1177/1479164112441190] [PMID: 22381403]

[96]   Abdul-Ghani MA, Norton L, DeFronzo RA. Efficacy and safety of SGLT2 inhibitors in the treatment of type 2 diabetes mellitus. Curr Diab Rep 2012; 12(3): 230-8.
[http://dx.doi.org/10.1007/s11892-012-0275-6] [PMID: 22528597]

[97]   Damkjær M, Isaksson GL, Stubbe J, Jensen BL, Assersen K, Bie P. Renal renin secretion as regulator of body fluid homeostasis. Pflugers Arch 2013; 465(1): 153-65.
[http://dx.doi.org/10.1007/s00424-012-1171-2] [PMID: 23096366]

[98]   Thomson SC, Rieg T, Miracle C, *et al.* Acute and chronic effects of SGLT2 blockade on glomerular and tubular function in the early diabetic rat. Am J Physiol Regul Integr Comp Physiol 2012; 302(1): R75-83.
[http://dx.doi.org/10.1152/ajpregu.00357.2011] [PMID: 21940401]

[99]   Holman RR, Paul SK, Bethel MA, Neil HA, Matthews DR. Long-term follow-up after tight control of blood pressure in type 2 diabetes. N Engl J Med 2008; 359(15): 1565-76.
[http://dx.doi.org/10.1056/NEJMoa0806359] [PMID: 18784091]

[100]   U.S. Food and Drug Administration. 2011.http://www.fda.gov/downloads/AdvisoryCommittees/ CommitteesMeetingMaterials/drugs/EndocrinologicandMetabolicDrugsAdvisoryCommittee/ucm2629 94.pdf

[101]   Bristol-Myers Squibb and AstraZeneca. 2011.http://www.fda.gov/downloads/AdvisoryCommittees/ CommitteesMeetingMaterials/Drugs/EndocrinologicandMetabolicDrugsAdvisoryCommittee/UCM262 996.pdf

[102]   Gallwitz B. The Cardiovascular Benefits Associated with the Use of Sodium-Glucose Cotransporter 2 Inhibitors - Real-World Data. Eur Endocrinol 2018; 14(1): 17-23.
[http://dx.doi.org/10.17925/EE.2018.14.1.17] [PMID: 29922347]

[103]   Zinman B, Wanner C, Lachin JM, *et al.* EMPA-REG OUTCOME Investigators. Empagliflozin, cardiovascular outcomes, and mortality in type 2 diabetes. N Engl J Med 2015; 373(22): 2117-28.
[http://dx.doi.org/10.1056/NEJMoa1504720] [PMID: 26378978]

[104]   Neal B, Perkovic V, Mahaffey KW, *et al.* CANVAS Program Collaborative Group. Canagliflozin and cardiovascular and renal events in type 2 diabetes. N Engl J Med 2017; 377(7): 644-57.
[http://dx.doi.org/10.1056/NEJMoa1611925] [PMID: 28605608]

[105]   Wiviott SD, Raz I, Bonaca MP, *et al.* DECLARE–TIMI 58 Investigators. Dapagliflozin and

Cardiovascular Outcomes in Type 2 Diabetes. N Engl J Med 2019; 380(4): 347-57.
[http://dx.doi.org/10.1056/NEJMoa1812389] [PMID: 30415602]

[106] Sattar N, McLaren J, Kristensen SL, Preiss D, McMurray JJ. SGLT2 Inhibition and cardiovascular events: why did EMPA-REG Outcomes surprise and what were the likely mechanisms? Diabetologia 2016; 59(7): 1333-9.
[http://dx.doi.org/10.1007/s00125-016-3956-x] [PMID: 27112340]

[107] Baartscheer A, Schumacher CA, Wüst RC, *et al.* Empagliflozin decreases myocardial cytoplasmic Na^+ through inhibition of the cardiac Na^+/H^+ exchanger in rats and rabbits. Diabetologia 2017; 60(3): 568-73.
[http://dx.doi.org/10.1007/s00125-016-4134-x] [PMID: 27752710]

SUBJECT INDEX

A

Atta-ur-Rahman (Ed.)